T0298326

Quantum Machine Learning

This book presents the research into and application of machine learning in quantum computation, known as quantum machine learning (QML). It presents a comparison of quantum machine learning, classical machine learning, and traditional programming, along with the usage of quantum computing, toward improving traditional machine learning algorithms through case studies.

In summary, the book:

- Covers the core and fundamental aspects of statistics, quantum learning, and quantum machines.
- Discusses the basics of machine learning, regression, supervised and unsupervised machine learning algorithms, and artificial neural networks.
- Elaborates upon quantum machine learning models, quantum machine learning approaches and quantum classification, and boosting.
- Introduces quantum evaluation models, deep quantum learning, ensembles, and QBoost.
- Presents case studies to demonstrate the efficiency of quantum mechanics in industrial aspects.

This reference text is primarily written for scholars and researchers working in the fields of computer science and engineering, information technology, electrical engineering, and electronics and communication engineering.

Quantum Machine Learning

A Modern Approach

Edited by

S. Karthikeyan
M. Akila
D. Sumathi
T. Poongodi

CRC Press is an imprint of the
Taylor & Francis Group, an informa business
A CHAPMAN & HALL BOOK

Designed cover image: ShutterStock

First edition published 2025
by CRC Press
2385 NW Executive Center Drive, Suite 320, Boca Raton, FL 33431

and by CRC Press
4 Park Square, Milton Park, Abingdon, Oxon, OX14 4RN

CRC Press is an imprint of the Taylor & Francis Group, LLC

Library of Congress Cataloging-in-Publication Data
Names: Karthikeyan, S. (Computer scientist), editor.
Title: Quantum machine learning : a modern approach / edited by S. Karthikeyan, M. Akila and D. Sumathi, T. Poongodi.
Other titles: Quantum machine learning (Karthikeyan) Description: First edition. | Boca Raton : C&H/CRC Press, 2025. | Includes bibliographical references and index. | Summary: “This book presents the research into and application of machine learning in quantum computation, known as quantum machine learning (QML)
Identifiers: LCCN 2024014463 (print) | LCCN 2024014464 (ebook) | ISBN 9781032544717 (hbk) | ISBN 9781032552323 (pbk) | ISBN 9781003429654 (ebk)
Subjects: LCSH: Machine learning. | Quantum computers. | Quantum theory.
Classification: LCC Q325.5 .Q36 2025 (print) | LCC Q325.5 (ebook) | DDC 006.3/1--dc23/eng/20240617
LC record available at https://lccn.loc.gov/2024014463
LC ebook record available at https://lccn.loc.gov/2024014464

ISBN: 978-1-032-54471-7 (hbk)
ISBN: 978-1-032-55232-3 (pbk)
ISBN: 978-1-003-42965-4 (ebk)

DOI: 10.1201/9781003429654

Typeset in Times
by SPi Technologies India Pvt Ltd (Straive)

Contents

PART III Quantum Models

PART IV Quantum Evaluation Models

Editors

Dr. Karthikeyan Saminathan is currently working as an Associate Professor (Research) in the Department of Cyber Security at Sri Venkateshwara College of Engineering, Bangalore, Karnataka, India and he is associated with Bluechip CyberTech Services as an Assistant Vice President which is the wing of Bluechip Services International Pvt Ltd, Bangalore. He is Founder CEO for AI Quantalytics Startup which primarily focuses on IT consulting and product development to AI-driven solutions and cloud innovations. His educational background includes receiving his B.E-CSE degree from Anna University, Chennai in 2010 and his M.E-Software Engineering from Anna University, Chennai in 2012. He received his PhD in Cloud and Bigdata Analytics from VIT University, AP. His research interests include artificial intelligence and machine learning, high-performance computing, cloud and big data analytics, and data sciences. He has published more than 60+ papers in reputed journals, seven book chapters, and more than 12 patents. He has delivered 100+ technical talks to various Academic Institutions and Industries. Also, he holds the prestigious role of NASSCOM Prime brand Ambassador. He is a life member of international professional bodies such as ISTE, IAENG, ISRD, and IFERP and he is also a senior member of IEEE.

Dr. M. Akila is a CEO of Metasage Alliance Consulting Expert Pvt Limited, Coimbatore, India. She is also an insatiable appetite for continuous learning and teaching, a philanthropic leader, a diligent researcher, and an experienced and insightful academician. With 27 years of experience, her mission is to enhance standards of education by providing an excellent, ingenious learning environment that is rational with the core values of Academic Institutions.

Dr. Akila's main research interest is machine learning with applications to computer vision and data science, but she is also interested in the efficient implementation of optimization algorithms in engineering problems. An inspiring, motivating, and committed CSE professor, she is actively engaged with professional organizations such as IET and IEEE and is an additional secretary in the Institution of Green Engineers. She has published three patents and delivered 32 invited lectures at various institutes. Dr. Akila received the Kalam 2020 award for service to Green Technology, 2018 and has been awarded the Cambridge International Certificate for Teachers and Trainers and a Certificate of Achievement from IGEN. She is, in addition, a renowned reviewer in neuro-computing, IET biometrics, IET image processing, and IET electronics letters.

Dr. D. Sumathi is currently serving as a professor Grade 1-SCOPE at VIT-AP University, Andhra Pradesh. She earned her BE degree in computer science and engineering from Bharathiar University in 1994 and her ME degree in computer science and engineering from Sathyabama University, Chennai, in 2006. She completed her doctoral degree at Anna University, Chennai. With a total of 23 years of experience, including 6 years in the industry and 17 years in the teaching field, she holds the additional responsibility of serving as Assistant Director of the Software Development Cell, which automates various campus upkeep functionalities.

Dr. Sumathi has taken on various administrative roles during her tenure. Her research interests encompass cloud computing, network security, data mining, natural language processing, machine learning, deep learning, and theoretical foundations of computer science. She has published numerous papers in reputed international journals and conferences. Furthermore, she has organized several international conferences, acting as a technical chair and tutorial presenter. Dr. Sumathi is a life member of ISTE and has published book chapters in CRC Press, IGI Global, Springer, IET, and edited books with publishers like CRC and Wiley. In addition to this, she holds patents related to the health sector. Her invited talks on domains such as machine learning, deep learning, and big data analytics have inspired budding researchers to engage in thoughtful research work and problem statements. Currently, she is guiding five research scholars within research areas in biomedical applications.

Prof. T. Poongodi is currently working as a Professor in the Department of Computer Science and Engineering at the Dayananda Sagar University, Bangalore, Karnataka, India. She received her PhD degree in information technology (information and communication engineering) from Anna University, Tamil Nadu, India. Her current research interests include network security, wireless ad hoc and sensor networks, internet of things (IoT), data science, and blockchain technology for emerging communication networks. She is CISCO, Oracle Academy, and Structured Query Language certified. Prof. Poongodi is the author of over 50 book chapters, including with reputed publishers such as Springer, Elsevier, IET, Wiley, De-Gruyter, CRC Press, IGI Global, and more than 30 international journals and conferences. She has published more than 15 authored/edited books in the areas of the internet of things, data analytics, blockchain technology, artificial intelligence, machine learning, and healthcare informatics, published by reputed publishers such as Springer, IET, Wiley, CRC Taylor & Francis, and Apple Academic Press. She adopts a universal and humanistic approach in her academic and research works. In her research, she has undertaken meticulous scientific studies of emerging issues in networking disciplines. She has more than 17 years of academic work experience in teaching and multidisciplinary research. She has received awards, namely the Research and Innovation Award (2019, 2020, 2021), and for Excellence in the areas of Research and Innovation/Academic Excellence/Extension Activities (2018–2019, 2019–2020) from Galgotias University. Prof. Poongodi has also received invitations to address international conferences as a keynote speaker. She is a reviewer for international journals and conferences and also has five Indian patents.

Contributors

Indhuja Anandan
SNS College of Technology
India

Reethika Anandan
Sri Ramakrishna Engineering College
India

Ramathilagam Arunagiri
PSR Engineering College
India

David Samuel Azariya
Sona College of Technology
India

Gomathy Balasubramanian
PSG Institute of Technology and Applied Research
Coimbatore, India

Soham S. Bhoir
KJ Somaiya College of Engineering
India

Mani Deepak Choudhry
SRM Institute of Science and Technology
Kattankulathur, India

Harshal H. Dave
KJ Somaiya College of Engineering
India

Nirmala Ganapathy
RMD Engineering College
India

Iniyan Shanmugam
SRM Institute of Science and Technology
India

Akshya Jothi
SRM Institute of Science and Technology
Kattankulathur, India

Sathya Karunanidhi
Vellore Institute of Technology
India

Akila Krishnamoorthy
SRM Institute of Science and Technology
Vadapalani, Chennai, India

Pranav Manikandan
SASTRA Deemed to be University
India

Sundarrajan Munusamy
SRM Institute of Science and Technology
Kattankulathur, India

Ganesh Kumar Natarajan
GITAM University
India

Ratna Kumari Neerukonda
SRM Institute of Science and Technology
India

Durgadevi Palani
SRM Institute of Science and Technology
Vadapalani, Chennai, India

Kanaga Priya Palanisamy
Sri Eshwar College of Engineering
India

Ponnuviji Namakkal Ponnusamy
RMK College of Engineering and Technology
India

Thiruselvan Palusamy
PSR Engineering College
India

Mangalraj Poobala
GITAM University
India

Prianka Ramachandran Radhabai
New Horizon College of Engineering
India

Ramani Ramasamy
PSR Engineering College
India

Devi Priya Rangasamy
KPR Institute of Engineering and Technology
India

Akshay Bhuvaneswari Ramakrishnan
SASTRA Deemed to be University
India

Anupama Cholanayakanahalli Govinda Reddy
SRM Institute of Science and Technology
India

Lalith Prem Ravi
Informatics Software and Hardware
India

Savitha Selvi Jagathiswaramoorthi
Sri Ramakrishna Institute of Technology
India

Karthikeyan Saminathan
MINE
Bengaluru, India
&
Sri Venkateshwara College of Engineering
Bengaluru, India

Thanga Revathi Shanmugakani
SRM Institute of Science and Technology
India

Jeevanandham Sivaraj
Sri Ramakrishna Engineering College
India

Shreyanth Srikanth
Birla Institute of Technology and Science
India

Nisha Soms
KPR Institute of Engineering and Technology
India

Indra Priyadharshini Sundar
VIT-Chennai
India

Hariharan Bagavathi Thevar
SRM Institute of Science and Technology
India

Justin Vargese
GITAM University
India

Siva Rathinavelayutham
SRM Institute of Science and Technology
India

Mohanraj Vijayakumar
Sona College of Technology
India

Part I

Introduction to Statistical and Quantum Learning

1 Fundamentals of Statistics

Mani Deepak Choudhry
SRM Institute of Science and Technology, India

Sundarrajan Munusamy
SRM Institute of Science and Technology, India

Jeevanandham Sivaraj
Sri Ramakrishna Engineering College, India

Akshya Jothi
SRM Institute of Science and Technology, India

1.1 STATISTICS AND ITS TYPES

1.1.1 Definition of Statistics

This chapter commences with an exploration of the fundamental concepts and significance of statistics. It delves into the reasons why the study of statistics is crucial in various fields. The subsequent sections encompass a range of essential topics that form the foundation of statistical analysis. These topics encompass key principles that are fundamental to understanding and conducting statistical research. Each section serves as a building block in the study of statistics, providing a comprehensive understanding of its principles and applications. The chapter aims to equip readers with the necessary knowledge to engage effectively in statistical analysis and make informed decisions based on data.

Statistics is a field within mathematics and a scientific discipline which encompasses the acquisition, examination, comprehension, demonstration, and arrangement of information, while ensuring complete originality and absence of plagiarism. It provides tools and techniques to make sense of large and complex datasets, enabling us to uncover patterns, trends, and relationships within the data. By employing statistical methods, we can draw reliable conclusions and make informed decisions in various fields, including business, economic, social sciences, and healthcare. The study of statistics incorporates mathematical calculations and numerical analysis, but equally it places substantial emphasis on the process of selecting appropriate data and interpreting the resulting statistics.

DOI: 10.1201/9781003429654-2

Statistics goes beyond the realm of raw data and numerical values, as it encompasses a wider spectrum of methodologies and processes aimed at evaluating, rendering, bestowing, and constructing well-informed judgments using the collected information. In its broadest sense, the term "statistics" refers to the entire framework that supports the understanding and utilization of data in various fields. Statistics provides a systematic approach to dealing with data, enabling researchers and analysts to extract meaningful insights from complex datasets. It involves employing statistical methods and tools to explore, summarize, and draw conclusions from data, ultimately facilitating evidence-based decision-making. The process of analyzing data entails the utilization of statistical methodologies to detect inherent outlines, leanings, and associations present within the dataset. These methodologies encompass a wide range of statistical metrics that aid in understanding and summarizing data. They include metrics of central tendency, such as the mean and median, which provide insights into the typical or central value of a dataset. Statistical models and algorithms are used to uncover underlying structures and associations within the data, allowing for deeper exploration and understanding. Interpreting data is a critical aspect of statistics, as it involves making sense of the statistical findings within the context of the research question or problem being investigated. This process requires domain knowledge, critical thinking, and a keen understanding of statistical concepts to derive meaningful insights and draw valid conclusions from the data. Displaying data is essential for effectively communicating statistical information. Visual representations, such as graphs, charts, and tables, are used to convey information in a vibrant and succinct way. These visualizations aid in the understanding and interpretation of complex statistical information, making it accessible to a wider audience. Making decisions based on data is the ultimate goal of statistical analysis. By using statistical methods to analyze and interpret data, decision-makers can gain valuable insights that inform their actions. Statistical techniques provide a framework for quantifying uncertainty, evaluating risks, and assessing the impact of various factors, allowing for more informed and evidence-based decision-making [1].

At its core, statistics involves the collection of data. This data can be in the form of numerical values, measurements, observations, or responses to surveys or experiments. The process of data collection is crucial to ensure the reliability and validity of statistical analyses. Various sampling techniques are employed to select representative samples from the complete set of interest, minimizing bias and ensuring generalizability of the results.

Data refers to factual information and numerical values that undergo collection, analysis, and summarization, aiding in the process of interpretation and presentation. The two broad categories of data are quantitative and qualitative. Quantitative data entails measurements pertaining to "how much" or "how many" of a particular attribute, whereas qualitative data involves assigning labels or names to categories of similar items [2–4].

As an illustration, let's examine a study that centers on characteristics like oldness, gender, wedded status, and yearly pay within a trial of 100 entities. These characteristics are referred to as variables, and each individual is linked to specific data values for each variable. For instance, an individual who is 28 years old, male, single, and earns an annual income of $30,000 would have statistical values of 28, masculine, solitary, and $30,000 logged. Considering there are 100 individuals and 4 variables, the dataset would consist of a total of 400 individual data items (100 × 4). Here, quantitative variables are oldness and yearly revenue. On the other hand, gender and wedded status are experiential variables.

By understanding the nature of quantitative and qualitative data, researchers can appropriately analyze and interpret the data, leading to valuable insights and informed decision-making.

Once the data is collected, statistical analysis techniques are employed to extract meaningful insights. Descriptive statistics summarize and describe the main features of data, including metrics of central tendency and metrics of dispersion. Inferential statistics, on the other hand, use trial facts to make implications or draw assumptions, allowing researchers to generalize findings beyond the immediate sample. These inferences involve estimating complete set constraints, testing theories, and calculating the level of uncertainty associated with the findings. This involves hypothesis testing, estimation, and determining the level of confidence in the results.

Statistical interpretation plays a critical role in comprehending the implications of an analysis. It involves deriving meaning from the statistical findings within the context of the research question or problem being addressed. This process enables researchers to draw meaningful conclusions, identify patterns or trends, and gain insights into the underlying relationships within the data. Statistical interpretation permits a deep consideration of the practical significance and real-world implications of the analysis [5–7]. This step requires critical thinking and domain knowledge to determine the significance and practical implications of the statistical results. It also involves considering potential limitations and sources of error in the analysis.

The final step in the statistical process is the presentation and organization of the findings. This includes the use of graphs, charts, tables, and reports to communicate the results effectively. Visual representations of data can aid in the understanding and interpretation of complex statistical information. Moreover, statistical software tools such as R, Python, and Excel facilitate the analysis and visualization of data, making statistical techniques accessible to a wider audience.

1.1.2 Importance of Statistics

The major importance of statistics and why it is needed are given below [8–10]:

- Statistics enables researchers and analysts to make sense of large and complex datasets.
- It provides a quantitative basis for decision-making and helps minimize uncertainty.
- Statistics helps identify patterns, trends, and relationships in data, leading to valuable insights.
- It aids in the design and implementation of effective research studies and experiments.
- Statistics plays a crucial role in evidence-based policy-making and program evaluation.

- It helps in the identification and understanding of complete set characteristics and behaviors.
- Statistics facilitates forecasting and prediction by analyzing historical data patterns.
- It provides a framework for testing hypotheses and drawing valid conclusions.
- Statistics supports risk assessment and management in various industries.
- It enables the comparison of data from different sources and contexts.
- Statistics helps in monitoring and tracking progress toward goals and targets.
- It provides a basis for sampling techniques, allowing researchers to generalize findings to larger complete sets.
- Statistics aids in identifying and addressing biases in data collection and analysis.
- It supports quality control and process improvement by analyzing production data.
- Statistics helps in resource allocation and optimization, improving efficiency.
- It plays a crucial role in financial analysis and investment decision-making.
- Statistics supports market research and helps businesses understand consumer behavior.
- It aids in the identification of trends and patterns in healthcare data for improved patient outcomes.
- Statistics is essential in social sciences to analyze social, economic, and demographic trends.
- It plays a crucial role in environmental research and policy-making.
- Statistics helps in analyzing survey data and public opinion research.
- It supports the identification and assessment of correlations between variables.
- Statistics aids in studying the effectiveness of interventions and treatments.
- It provides a foundation for statistical software development and data analysis tools.
- Statistics enables data-driven decision-making, leading to improved efficiency and effectiveness in various domains.

1.1.3 Basic Terminology in Statistics

- ***Complete set:*** A complete set, in the context of statistics, refers to the entire set of individuals, objects, or elements from which data can be collected and analyzed. It represents the complete group that is of interest to the researcher or analyst. The complete set can vary in size, ranging from small groups to large communities, and can encompass diverse entities such as people, animals, plants, or inanimate objects.
- *For example*: Presume a scholar needs to investigate the eating habits of teenagers in a particular city. The complete set in this case would be all the teenagers living in that city. By accurately defining the complete set

of interest and carefully selecting a representative sample, researchers can gather insights and draw meaningful conclusions that can inform public health initiatives, nutritional education programs, or policy decisions aimed at improving the eating habits of teenagers in that city.

- ***Sample:*** A portion of a complete set that is deliberately chosen and utilized in data sampling and inferential statistics to make predictions and draw valid conclusions pertaining to the entire complete set is referred as a "sample".
- *For example*: Let's consider an example to illustrate the concept of a sample in relation to a complete set. Imagine a researcher is interested in studying the typical height of students in a university. The populace in this case would be all the students enrolled in that university. In this example, the populace details all students in the university, while the trial represents a subset of 200 randomly selected students. By studying the sample, the researcher can make predictions and draw conclusions about the average height of the entire student complete set, providing insights without the need for a whole complete set measurement.
- ***Variable:*** A variable, in the realm of statistics, is a value, characteristic, or extent that can be constrained or calculated. It represents an attribute or property that can vary among individuals, objects, or events within a complete set or sample. Variables can take on different forms, such as numerical values or categorical labels, and they serve as the building blocks for data analysis and statistical investigation. Each variable represents a data point that contributes to the overall understanding and analysis of a particular phenomenon or research question.
- *For example*: To illustrate the concept of a variable, Let's explore an example of a survey undertaken to examine the correlation between the duration of exercise and heart health. In this scenario, two variables of interest would be "hours of exercise per week" and "heart health status." The variable "hours of exercise per week" denotes a quantifiable and measurable numerical value. In this example, the variables "hours of exercise per week" and "heart health status" represent distinct characteristics that can vary among individuals in the sample. These variables serve as data points that contribute to understanding the association between exercise habits and heart health outcomes.
- ***Probability distribution:*** A fundamental numerical concept used in statistics to quantify the likelihood or probabilities associated with different possible outcomes of an experiment or random event is "probability distribution". It provides a systematic and structured framework for understanding the relative chances or probabilities of different outcomes occurring.
- ***Statistical parameter:*** A statistical or complete set parameter refers to a numerical quantity that serves as an index or characteristic of a specific complete set or probability distribution. These parameters provide valuable insights into the distribution's central tendencies, shape, and other important characteristics. These parameters capture important aspects of the complete set's distribution and help summarize its overall behavior.

1.1.4 Functions and Scope of Statistics

Statistics is a well-defined branch of research that encompasses the development and application of techniques to effectively collect, organize, present, analyze, and interpret data. Its primary objective is to assess the reliability of conclusions through the use of probability statements [11].

By utilizing statistical methods and processes, businesses can harness the power of vast numerical facts. These methods enable businesses to uncover the underlying stories and insights hidden within the data. Each figure represents more than just a numerical value; it carries a narrative that can inform decision-making, drive strategic planning, and fuel business development.

The role of statistics in business is crucial as it provides a systematic framework to extract meaningful information from data. This involves various stages, starting from the collection and organization of data, followed by its presentation in a format that facilitates analysis. Statistical techniques are then employed to uncover patterns, relationships, and trends within the data, leading to valuable insights and actionable knowledge.

The interpretation of statistical findings is vital for businesses as it comprises consideration of implications and practical consequences of the outcomes. This requires skilled analysts who can critically evaluate the statistical outcomes in the context of business objectives and industry dynamics. By analyzing and interpreting the data, businesses can make informed decisions, identify areas for improvement, and devise strategies to optimize their operations.

The reliability of statistical conclusions is a fundamental aspect that sets statistics apart as a scientific discipline. The use of probability statements allows businesses to quantify the uncertainty associated with the conclusions drawn from the data. This approach provides a measure of confidence and allows decision-makers to gauge the reliability of the insights derived from statistical analyses.

1.1.4.1 Key Functions of Statistics

Some of the key functions of statistics are outlined below.

1. Condensation
    - Statistics offers a valuable tool for compressing large volumes of data into concise and meaningful information.
    - By utilizing statistical techniques, various types of data, such as aggregated sales forecasts, stock market indices like BSE, or GDP growth rates, can be summarized effectively.
    - This compression enables decision-makers to grasp essential insights and make informed judgments in a more manageable and efficient manner.
    - The advantage of using financial ratios lies in their simplicity and interpretability. They serve as meaningful benchmarks for assessing a company's financial health and performance.
    - Decision-makers can easily compare ratios against industry standards, historical trends, or competitor benchmarks to gain valuable insights.
    - Financial ratios enable quick comparisons and evaluations, saving time and providing a clear snapshot of a company's profitability and financial standing.

- While financial ratios provide valuable insights, they should be considered alongside other contextual factors and a comprehensive understanding of the business dynamics for informed decision-making.

2. Comparison
   - Statistics plays a crucial role in facilitating the comparison of different quantities, allowing for meaningful assessments and informed decision-making.
   - By utilizing statistical methods, researchers and analysts can establish frameworks for comparing variables, groups, or complete sets, enabling a deeper understanding of relationships, differences, and similarities. For illustration, consider a study comparing the efficacy of two different marketing approaches in promoting a product. By employing statistical techniques, researchers can collect data on key metrics such as sales revenue, customer engagement, or brand awareness for each strategy. They can then analyze the data using appropriate statistical tests to determine if one strategy outperforms the other or if there are significant differences in their effectiveness.
   - Furthermore, statistics allows for the comparison of different groups or complete sets. For example, a scholar may be concerned with comparing the average income levels of employees across various industries. By collecting income data from representative samples of employees in each industry, statistical analysis can be applied to regulate if there are noteworthy modifications in average incomes between the groups.
   - Comparisons in statistics are not limited to numerical quantities alone. Categorical variables can also be compared using statistical methods. For instance, researchers may examine the association between gender (a categorical variable) and the likelihood of purchasing a particular product.
   - Statistics serves as a powerful tool for comparing different quantities, whether they are numerical values, groups, or categorical variables.
   - Through statistical analysis, researchers can assess the significance of differences, identify relationships, and make meaningful comparisons that contribute to a deeper understanding of the data.
   - These comparisons support evidence-based decision-making in various fields, ranging from marketing and business to healthcare and social sciences.
3. Forecasting
   - Statistics plays a critical role in forecasting by examining the trends and patterns of variables. It is an essential tool for effective planning and decision-making. Relying on intuition or guesswork for predictions can lead to disastrous outcomes for businesses, highlighting the significance of statistical forecasting.
   - To illustrate, let's consider the scenario of deciding the production capacity for a vehicle-manufacturing plant. Before making significant investments, it is crucial to predict the future demand for the product mix, availability of components, the cost of manpower, and competitor strategies over the next five to ten years.
   - By employing statistical forecasting techniques, researchers and analysts can analyze historical data on variables such as sales volumes, market

trends, economic indicators, and customer preferences. These techniques can identify underlying patterns, seasonality, and trends in the data, providing valuable insights into future demand and market conditions.

- For instance, time series analysis can be utilized to analyze historical sales data and identify patterns, such as recurring seasonal fluctuations or long-term trends. These patterns can be used to forecast future demand for different product variants or segments. Additionally, regression analysis can help assess the impact of various factors, such as competitor strategies or economic indicators, on the demand for vehicles.
- By integrating multiple sources of data and applying statistical models, researchers can generate forecasts that inform decisions regarding production capacity. These forecasts enable businesses to align their resources, workforce, and production plans to meet anticipated demand and optimize their operations.
- The utilization of statistical forecasting minimizes the risks associated with inaccurate predictions based on intuition or subjective opinions. By relying on data-driven insights, businesses can make informed decisions and allocate resources efficiently. Statistical forecasting empowers organizations to adapt to changing market conditions, anticipate customer needs, and stay ahead of competitors.

4. Testing of hypotheses
   - Hypotheses are essential statements made by researchers regarding complete set parameters, which are based on existing knowledge derived from literature. These hypotheses serve as propositions that researchers aim to test for validity and examine in light of new information or data. By formulating hypotheses, researchers can investigate specific relationships, effects, or differences within complete sets.
   - When drawing inferences about the complete set based on sample estimates, there is an inherent element of risk involved. This is because the sample, although representative of the complete set, may not capture the entire variability or characteristics present in the wider population. In other words, the sample may not perfectly reflect the true values of the complete set parameters.
   - During the process of hypothesis testing, researchers collect and analyze sample data to draw assumptions about the complete set. Hypothesis testing utilizes statistical techniques to assess the probability of perceiving the attained trial outcomes as assumption of true for zero hypothesis. This helps analysts to evaluate the validity of the hypothesis and make informed decisions regarding its acceptance or rejection.
   - However, due to the limited size and scope of the sample, there is always a risk of making errors in the inference process.
   - To minimize these potential risks, researchers employ strategies such as confidence intervals and p-values.
   - It is important to recognize that hypothesis testing and drawing inferences from samples involve a balance between accepting the inherent risk and ensuring the validity of the conclusions. Researchers strive to minimize errors and uncertainties by designing studies carefully, using

appropriate statistical methods, and interpreting the results within the appropriate context.

5. Precision
   - Statistics serves as a powerful tool to visualize and present facts in a precise and quantitative form. It enables researchers and analysts to convey information using numerical data, which often carries greater weight and conviction compared to qualitative data. Quantitative information provides a clear and objective representation of facts, enhancing the credibility and persuasiveness of the presented findings.
   - For illustration, consider a study probing the usefulness of a new drug in treating a specific medical condition. By conducting a quantitative analysis of clinical trial data, researchers can present the results in a statistical format. They might showcase the percentage of patients experiencing symptom improvement, the reduction in the average duration of symptoms, or the statistical significance of the drug's efficacy compared to a placebo. These quantitative measures provide concrete evidence of the drug's effectiveness and can be more persuasive in influencing medical practitioners or policymakers to adopt the treatment.
   - When information is presented in a quantitative form, it offers numerous advantages when related to qualitative data. Primarily, quantitative data allows for more precise comparisons and measurements. For instance, instead of describing a product as "good" or "bad", numerical ratings or performance metrics can be used to convey its quality or effectiveness objectively. These quantifiable measures provide a standardized and precise way of understanding and comparing different entities or phenomena.
   - In addition, quantitative data facilitates data-driven decision-making. The use of statistical analysis enables researchers to identify patterns, trends, and relationships within the data. By quantifying these insights, decision-makers can make informed judgments based on objective evidence rather than relying solely on subjective opinions or qualitative observations.
   - Moreover, the use of statistics enhances the reproducibility and transparency of research. Quantitative data allows other researchers to replicate and validate findings, increasing the reliability and trustworthiness of the information. Statistical analyses provide a framework for hypothesis testing, significance determination, and the reporting of confidence intervals, all of which contribute to the robustness and replicability of research findings.
   - However, it is crucial to note that the value of quantitative information does not undermine the importance of qualitative data in certain contexts. Qualitative data, such as interviews or open-ended survey responses, can provide valuable insights into individuals' perspectives, experiences, or motivations. These qualitative insights can complement and enrich the quantitative findings, offering a more comprehensive understanding of complex phenomena.
6. Expectation
   - Statistics serves as a fundamental building block for formulating clear plans and policies. It provides the necessary framework to make informed

decisions by assessing the expected outcomes under various circumstances. For instance, determining the quantity of raw material to import in a year, the extent of capacity expansion, or the number of recruits needed, relies on statistical analysis and forecasting.

- Let's consider an example to illustrate this concept. Imagine a manufacturing company that produces electronic devices. The management team wants to develop a plan for raw material procurement and production capacity to meet the expected demand over the next year. By employing statistical techniques, analysts can examine past sales data, evaluate market leanings, and consider other pertinent factors to make predictions about forthcoming claim.
- Based on the statistical analysis, the team can estimate the expected value of the outcome under different scenarios. They can assess the potential demand fluctuations, seasonal patterns, and market dynamics to determine the appropriate quantity of raw material to import, considering factors like lead times, production timelines, and inventory management.
- Similarly, statistical analysis can help in decision-making related to capacity expansion. By examining historical production data and analyzing trends, the management team can identify patterns of growth and evaluate the potential need for increased capacity. Statistical forecasting techniques can be employed to estimate the expected outcomes under different expansion scenarios, considering factors such as market demand, market share targets, and production efficiency.
- Furthermore, statistical models can aid in determining the required workforce based on projected production volumes and other operational considerations. By analyzing historical production data, workload patterns, and productivity metrics, the team can estimate the number of employees needed to meet the production targets while considering factors such as labor costs, skill requirements, and shifts.
- In all these scenarios, statistics provides the necessary tools to analyze data, make predictions, and assess the expected outcomes of various decisions. By employing statistical techniques, organizations can develop clear plans and policies that are based on evidence and informed by quantitative analysis. This enables businesses to optimize their operations, allocate resources efficiently, and adapt to changing market conditions.

1.1.4.2 Scope of Statistics

Significant capabilities within the scope of statistics are discussed below. Statistics can:

1. Present facts in numerical figures
   - Projecting a given problem in numerical form is a key principle of statistics.
   - By converting a problem or situation into numerical figures, statistics provides a structured and quantitative framework for better understanding the nature and characteristics of the problem at hand.

- This numerical presentation offers several advantages in terms of clarity, precision, and objectivity.

2. Present complex facts in a simplified form
   - Statistics has the ability to present complex facts and information in a shortened and effortlessly comprehensible way.
   - Statistics play a crucial role in exploring and investigating the relationship between different phenomena, such as income and consumption, demand and supply, and many others, allowing researchers to uncover valuable insights and understand the underlying connections.
3. Help in the formulation of policy
   - The utilization of statistical analysis in various economic, business, and governmental activities serves as the foundation for policy formulation.
   - Through statistical techniques, organizations can gain insights into consumer tastes and preferences, enabling them to align their product offerings accordingly and make informed decisions to meet market demands.

1.1.5 Types of Statistics

The common categories of statistics shown in Figure 1.1 are described in more detail below.

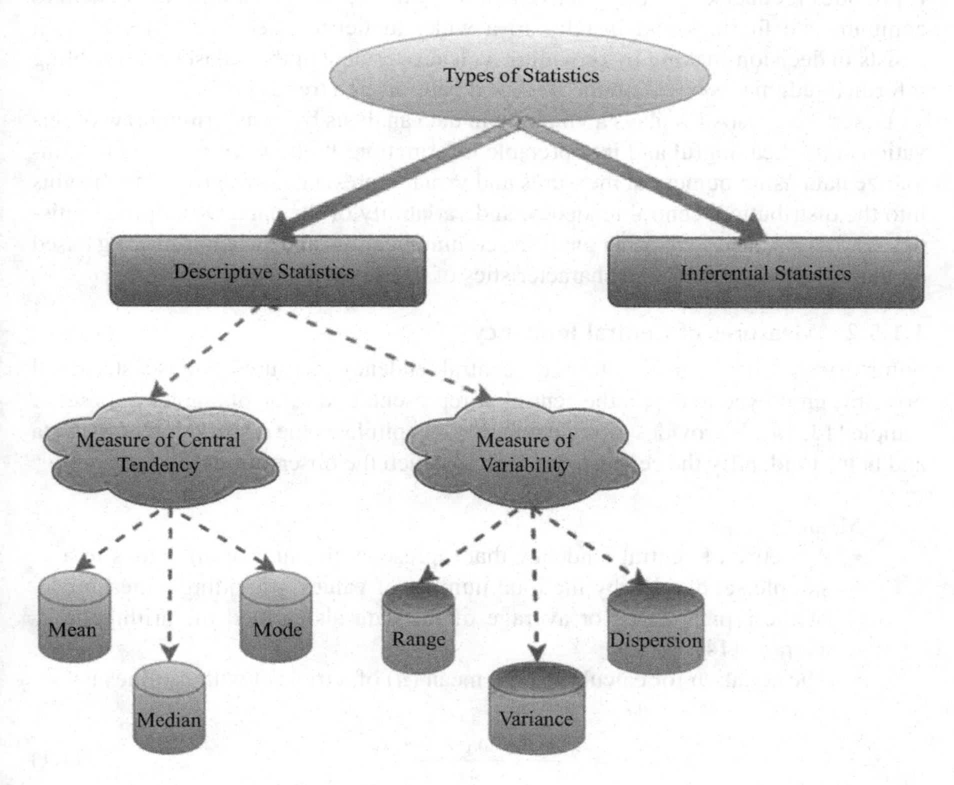

FIGURE 1.1 Types of statistics.

1.1.5.1 Descriptive Statistics

Descriptive statistics is a fundamental concept that enables the analysis, summarization, and organization of data through numerical measures and graphical representations. It provides a systematic approach to convert raw observations into meaningful information that can be easily understood and interpreted [12].

Through descriptive statistics, data is transformed into various formats such as numbers, graphs, bar plots, histograms, and pie charts. These visual representations provide researchers and analysts with the opportunity to acquire valuable insights into the distribution, patterns, and distinctive attributes of the data.

Measures like standard deviation and central tendency are commonly used in descriptive statistics. Central tendency measures summarize the typical or central value of the data, helping to identify a representative value around which the observations are centered.

By utilizing descriptive statistics, researchers can gain a comprehensive understanding of their existing data. It aids in the process of data exploration and discovery, allowing for the identification of outliers, skewness, or other notable characteristics within the dataset. Furthermore, it facilitates the comparison of data across different groups or categories, enabling researchers to identify similarities or differences.

Descriptive statistics serves as a foundation for further analysis and interpretation. It provides a concise and informative summary of the data, enabling researchers to communicate findings and insights to a wider audience effectively. Moreover, it assists in decision-making by providing a clear overview of the dataset and enabling informed judgments based on the observed patterns and trends [12].

Descriptive statistics plays a vital role in data analysis by transforming raw observations into meaningful and interpretable information. It allows researchers to summarize data using numerical measures and visual representations, providing insights into the distribution, central tendency, and variability of the data. Descriptive statistics serves as a basis for further analysis, communication, and decision-making based on the observed patterns and characteristics of the dataset.

1.1.5.2 Measures of Central Tendency

Summary statistics is another term for central tendency measures, and is a statistical principle employed to depict the central or representative value of a given data set or sample [13, 14]. It provides a single rate that recapitulates the distribution of the data and helps to identify the center point around which the observations cluster.

1. Mean
    - A metric of central tendency that represents the sum of all values in a sample set divided by the total number of values, providing a measure of the typical value or average of the data also called the arithmetic average [14].
    - The equation for calculating the mean (μ) of a trial set with n values is:

$$\mu = \frac{x_1 + x_2 + x_3 + \cdots + x_n}{n} \tag{1.1}$$

- Equation (1.1) gives the mean calculation where $x_1 + x_2 + x_3 + \cdots + x_n$ represent the individual values in the sample set, n represents the set count of values, and μ is the mean.

2. Median
   - The median serves as a measure of central value in a sample set and is determined by arranging the data set in ascending order and identifying the exact middle value. It represents the value that separates the higher half from the lower half of the ordered data set [14].
   - The equation for finding the median of a sample set with an odd number of values is:

$$M = \frac{n+1}{2} \tag{1.2}$$

   where n represents the sample set count.
   - The equation for an even number of data sets is:

$$M = \frac{\left(x\left[\frac{n}{2}\right] + x\left[\left(\frac{n}{2}\right) + 1\right]\right)}{2} \tag{1.3}$$

   where x represents the individual values in the sample set.

3. Mode
   - The mode represents the value that appears most commonly in a sample set. It corresponds to the data that is repeated the greatest number of times, highlighting the most common observation in the central set of data [14]. For example:

$$\left[2,3,4,2,4,6,4,7,7,4,2,4\right] \rightarrow \text{mode is } 4$$

1.1.5.3 Degree of Variability

The degree of variability, also referred to as the degree of dispersion, is a statistical principle employed to characterize the extent or spread of data within a sample or complete set. Within statistics, the range, variance, and standard deviation are three frequently utilized measures of variability.

1. Range
   - The range quantifies the degree of dispersion among values within a sample or data set. It is computed by determining the variation among the high and low values in the set, thereby representing the overall span or dispersion of the data.

$$\text{Range} = \text{maximum value} - \text{minimum value} \tag{1.4}$$

2. Variance
   - Variance, as a statistical measure, characterizes the degree to which a random variable deviates from its anticipated value. It is calculated by averaging the squared differences between each value and the expected value, offering a numerical measure of the overall variability or dispersion exhibited by the data.

$$S^2 = \sum_{i=1}^{n} \left[\left(x_i - \underline{x} \right)^2 \div n \right] \quad (1.5)$$

In this formula, n is data points count, $\bar{x}$ is mean, and x_i is distinct data point.

1.1.5.4 Inferential Statistics

Inferential statistics encompasses the process of drawing conclusions and making predictions about an entire complete set by utilizing a sample of data collected from that complete set.

- It extends the findings from a smaller sample to a larger complete set by utilizing probability theory to draw meaningful conclusions.
- Inferential statistics serves as a complementary tool to descriptive statistics, helping analyze and interpret results in order to make informed decisions and draw meaningful conclusions [15].
- One of its key applications is hypothesis testing, where the main objective is to evaluate and potentially reject the null hypothesis.

The process of conducting inferential statistics involves several essential steps, outlined below, to draw meaningful conclusions and make predictions:

1. The process begins with obtaining a theory that serves as the foundation for the research. From there, a research hypothesis is generated, stating the expected relationship or difference between variables. The variables are operationalized or defined in measurable terms to facilitate data collection and analysis.
2. Subsequently, the identification or determination of the complete set to which the study findings can be generalized is undertaken. Based on this complete set, a null hypothesis is formulated, which assumes no significant relationship or difference between the variables. A sample is then collected from the complete set, typically using appropriate sampling methods, and the study is conducted.
3. After data collection, statistical tests are performed to assess whether the characteristics observed in the illustration are significantly distinct from what would be probable under the null hypothesis. These trials help clarify the presence of meaningful relationships or differences and provide evidence to either discard or be unsuccessful to cast-off the null hypothesis.

4. The goal of conducting statistical tests is to assess the probability of gaining the experimental results solely by chance. Through comparing the obtained outcomes with the anticipated results under the null hypothesis, researchers can ascertain the degree of statistical significance and draw well-founded conclusions regarding the research hypothesis.

Throughout the process, it is important to maintain methodological rigor, account for biases, and use appropriate statistical techniques to guarantee the legitimacy and dependability of the findings.

Types of inferential statistics include:

- Confidence Intervals: Range
- Hypothesis Testing: Significance
- Regression Analysis: Prediction
- Analysis of Variance (ANOVA): Comparison
- Chi-Square Test: Association
- T-tests: Difference
- Correlation Analysis: Relationship

1.2 TYPES OF DATA

In the field of statistics, data can be categorized into distinct types, according to their inherent nature and defining characteristics [16]. Some widely acknowledged types of data include:

- **Nominal data:** Nominal data refers to categorical variables that symbolize distinct categories or groups, lacking any inherent numerical value or order. Instances of nominal data encompass stereotype, marital status, or types of cars (sedan/SUV/hatchback).
- **Ordinal data:** Ordinal data, similar to nominal data, represent categories or groups; however, they possess an inherent order or ranking within these categories. This means that the values can be ranked or ordered in a meaningful way. For instance, an ordinal variable could include rankings such as "high", "medium", and "low" or ratings like "strongly agree", "agree", "neutral", "disagree", and "strongly disagree".
- **Interval data:** These are numerical variables wherein the disparity between values holds significance, yet no absolute zero point exists. These data can be subjected to addition, subtraction, and averaging operations. Examples of interval data include temperature measurements in Celsius or Fahrenheit, as well as years denoted by numerical values (e.g., 2000, 2001, 2002).
- **Ratio data:** Ratio data are numerical variables that possess a true zero point, representing an absence of the measured attribute. These data allow for all arithmetic operations, including multiplication and division. Examples include weight, height, age, or income.
- **Discrete data:** Discrete data are values that can only take specific, separate, and distinct numerical values. They usually arise from counting or enumeration

processes and do not have fractional or intermediate values. Examples of interval data include the count of children in a family, or the number of cars present in a parking lot.
- **Continuous data:** These can take any numerical value within a given range. They can include fractional or decimal values, providing a high level of precision and flexibility in measurement. Examples include height, weight, or time.

Understanding the type of data is crucial as it determines the appropriate statistical techniques and methods for analysis. Different types of data require different approaches for summarizing, analyzing, and drawing meaningful conclusions.

1.3 COMPLETE SET MEASURE

In statistics, a complete set denotes the complete assemblage or entirety of individuals, objects, or events that capture the attention of a researcher and are of interest for study purposes [12].

- The complete set is the complete set of units that share a common characteristic or attribute of interest.
- It can vary depending on the research context and can be as specific as a group of students in a particular school or as broad as all the people living in a country.
- The concept of a complete set is fundamental in statistics as it provides the basis for making inferences and generalizations about a larger group based on information collected from a smaller subset called a sample.
- In studying the features of the populace, the researchers' purpose is to gain insights and draw conclusions about the broader target group.
- Complete sets can be finite or infinite. A **finite complete set** consists of a fixed number of elements, such as the students counted in a classroom, or cars produced by a manufacturer in a given year. An **infinite complete set**, on the other hand, has an unlimited or unknown number of elements, such as the heights of all individuals in a country.
- Complete set measures, also known as complete set parameters, are numerical values that describe specific characteristics of a complete set in statistics. These measures offer a condensed representation of the complete set and are employed to draw conclusions and make inferences about the complete set, utilizing data collected from a sample. Some commonly used complete set measures include:
    - ***Complete set mean*** **(μ):** The arithmetic average of a specific variable across the entirety of the complete set. It represents the central tendency of the complete set.
    - ***Complete set median*****:** The central value within the ordered sequence of observations in the complete set. It serves as the central value that distinguishes the upper half from the lower half of the complete set.

- ***Complete set mode*****:** This represents the value in a complete set or dataset that occurs with the peak frequency. It corresponds to the value that has the greatest occurrence or frequency among the data points.
- ***Complete set variance* (σ^2):** This quantifies the average squared deviation of individual information facts from the complete set mean, offering a measure of the complete set's dispersion or spread.
- ***Complete set standard deviation* (σ):** $\sqrt{\sigma^2}$ serves to quantify the average deviation of data points from the complete set mean.
- ***Complete set proportion* (σ):** Complete set proportion is a measure of the relative frequency of a specific category or attribute in the complete set. It is expressed as a ratio or percentage.

These complete set measures provide valuable insights into the characteristics and distribution of the complete set. However, it is often challenging or impractical to obtain the true complete set measures directly. Instead, statistical techniques are used to estimate these parameters based on information collected from a sample.

1.4 SAMPLING

Sampling is a valuable method in statistics that enables researchers to gather information by studying a representative subset, or trial, without the need to examine each individual unit. This approach provides a practical and efficient way to draw conclusions and make inferences about the larger complete set while minimizing costs and time constraints [12].

Figure 1.2 illustrates the process of sampling. To better grasp the concept of sampling, let's consider an example that illustrates its intuitive nature. Imagine a company that wants to assess the job satisfaction of its employees. Instead of surveying every employee, which may be time-consuming and costly, they decide to select a random trial of employees representing different departments and positions. By collecting responses from this sample, they can infer the overall job satisfaction of the entire workforce with a reasonable level of confidence, without the need to survey each and

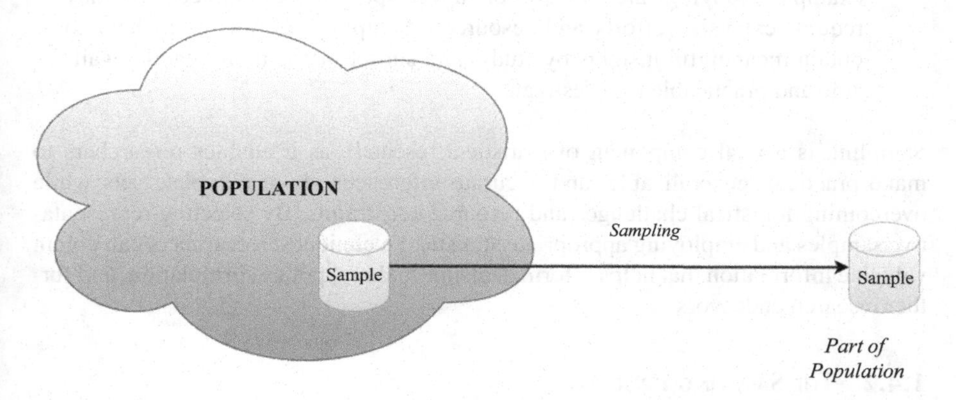

FIGURE 1.2 The sampling process.

every employee individually. This example demonstrates how sampling allows us to gain insights about a complete set by studying a subset, making data collection more feasible and providing valuable information for decision-making processes.

1.4.1 Reasons for Sampling

- **Practicality and efficiency:** When the complete set of interest is large or inaccessible, it may be impractical or impossible to study every individual or element. Sampling allows researchers to obtain representative information about the complete set by studying a smaller subset, making data collection more feasible, efficient, and cost-effective.
- **Generalizability:** A well-designed and representative sample can provide reliable estimates and implications about the larger complete set. By ensuring that the sample accurately signifies the characteristics of the complete set, researchers can generalize their findings and make valid conclusions about the entire target group.
- **Time and resource constraints:** Conducting a comprehensive study of an entire complete set may require an excessive amount of time, effort, and resources. Sampling enables researchers to collect an adequate amount of data within viable timeframes and budget constraints, rendering research more feasible, practical, and attainable.
- **Precision and accuracy:** Sampling, when properly executed, can yield precise and accurate results. By using statistical techniques, researchers can estimate the degree of uncertainty associated with the sample estimates.
- **Ethical considerations:** In some cases, studying the entire complete set may raise ethical concerns, such as invasion of privacy or exposing individuals to potential harm. Sampling provides a way to respect privacy and minimize potential risks by gathering information from a smaller group while still maintaining scientific rigor.
- **Accessibility and feasibility of data collection:** Certain complete sets or phenomena may be challenging to access or study comprehensively. For example, studying rare diseases or highly specialized complete sets may require extensive efforts and resources. Sampling allows researchers to obtain meaningful insights by studying a subset that is more readily available and practicable to investigate.

Sampling is a vital component of statistical research, as it enables researchers to make practical, generalizable, and accurate inferences about complete sets while overcoming logistical challenges and resource constraints. By selecting representative samples and employing appropriate statistical techniques, researchers can obtain valuable information that helps inform decision-making, policy formulation, and further research endeavors.

1.4.2 The Sampling Process

Figure 1.3 illustrates the steps involved in the process of sampling.

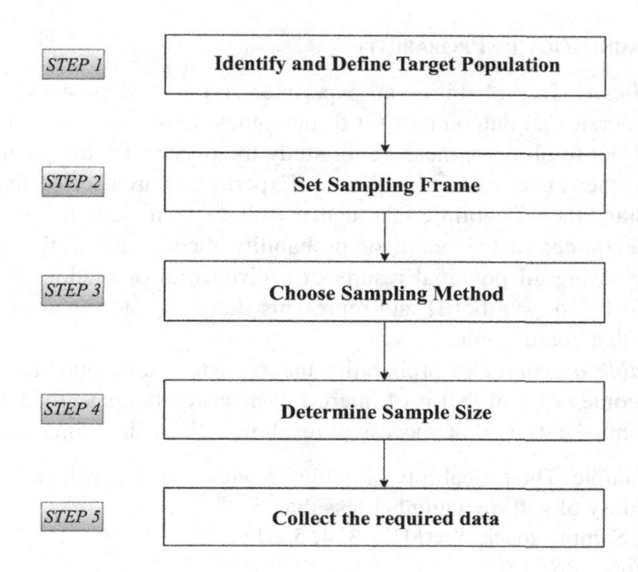

FIGURE 1.3 Steps involved in sampling.

1.5 CORRELATION BETWEEN VARIABLES

Correlation between variables is a statistical metric that provides a quantitative assessment of the interconnection between two variables. This metric is particularly useful when the variables exhibit a linear association with each other. To visually represent the data's fit, a scatterplot can be employed. By examining the scatterplot, we can gain insights into the relationship among the variables and determine the interrelation. This process allows us to assess the strength and direction of the correlation, enabling a deeper understanding of the connection between the variables.

1.6 PROBABILITY

In statistics, probability is a fundamental concept that evaluates the probability or likelihood of an event taking place. It quantifies the level of uncertainty connected to the result of an experiment or observation. Probability is used to analyze and understand the random nature of data and provides a mathematical framework for drawing conclusions on the value [17].

Probability is commonly expressed as a numerical value ranging from 0 to 1. A probability of 0 signifies an impossible event, while a probability of 1 denotes a certain event. The probability between 0 and 1 reflects the likelihood of occurrence for a given event.

Probability plays a crucial role in various statistical analyses, such as hypothesis testing, confidence intervals, and regression analysis [18]. It helps in making informed decisions, modeling uncertainties, and assessing the reliability of statistical results.

Equation 1.6 gives the event probability formula:

$$\text{Probability}\left(\text{Event}\right) = \frac{\text{Favorable outcomes}\left(x\right)}{\text{Total outcomes}\left(n\right)} \tag{1.6}$$

1.6.1 Terminology in Probability

- ***Experiment*:** In probability, an experiment refers to a process or activity that generates an outcome or set of outcomes. It is a controlled procedure carried out to observe, measure, or study the occurrence of specific events or outcomes under certain conditions. Experiments are used to analyze and understand the probabilities associated with different outcomes.
- ***Sample space*:** In the realm of probability theory, this is the collection encompassing all potential results of a given trial or random event. It is denoted by the symbol Ω and represents the complete range of potential results that could occur.
- ***Favorable outcome*:** In probability theory, a favorable outcome refers to an outcome or event that is of interest or meets certain specified criteria. It represents the desired or successful result in a given situation.

For example: The probability of getting 5 when a dice is rolled.
Probability of getting a number less than 5
Given: Sample space, $S = \{1, 2, 3, 4, 5, 6\}$
Therefore, $n(S) = 6$
Let A be the event of getting a number less than 5.
Then, $A = \{1, 2, 3, 4\}$
So, $n(A) = 4$
Using the probability equation:

$$p(A) = (n(A))/(n(S))$$

$$p(A) = 4/6 \tag{1.7}$$

$$m = 2/3$$

The probability of getting a number less than 5 is 2/3.

1.7 PROBABILITY DISTRIBUTIONS

In statistics, a probability distribution pertains to a mathematical function or model used to depict the probability of various outcomes or events transpiring within a specified dataset or complete set. It offers a structured approach to assigning probabilities to every potential value or range of values that a random variable can assume.

A probability distribution summarizes the probabilities of various outcomes and provides insights into the relative frequencies or likelihoods of those outcomes. It is a fundamental tool in statistical analysis, allowing researchers to quantify uncertainties, make predictions, and draw conclusions based on observed data [19].

1.7.1 Types of Distributions

1. **Discrete probability distribution:** This is applicable when the random variable can only take a finite count of distinct points or a countable number of values. It assigns probabilities to each possible value, forming a

probability mass function (PMF). Examples of discrete probability distributions include *binomial distribution, Poisson distribution*, and *geometric distribution.*

- *Binomial distribution*: This finds widespread application across diverse fields as a model for outcomes involving the count of successes or failures within a predetermined number of trials. Examples include the heads count gained in a sequence of coin flips or the count of defective items within a batch of products. It is crucial to acknowledge that binomial distribution relies on the assumptions of trial independence. Adherence to these assumptions is vital for the accurate utilization of binomial distribution.
- *Poisson distribution*: This is commonly used in various fields to model events such as the number of reception of calls in a phone booth within a specific time period, the number of accidents in a given day, or count of arrivals at a service point. It is important to note that Poisson distribution assumes independence among events and a constant rate of occurrence throughout the interval. Additionally, it is most accurate when the average rate of events is low, and the events are rare. In essence, Poisson distribution serves as a probability distribution for the count of events transpiring within a predetermined interval. Poisson distribution is particularly useful for modeling rare events or situations where events occur randomly and independently.
- *Geometric distribution*: This is commonly used in various fields to model events such as the coin flips count needed to obtain the first head, the number of attempts required to make a successful sale, or the number of failures before achieving a desired outcome. It is important to note that geometric distribution assumes independence among trials and a constant probability of success throughout the trials. Additionally, it is assumed that the trials are repeated until the first success occurs. To summarize, geometric distribution is particularly useful for modelling situations where success is achieved after a varying number of trials.

2. **Continuous probability distribution:** This is applicable when a random variable has the potential to assume any value within a particular interval or range. It is distinguished by a probability density function (PDF), which represents the likelihood of the variable taking on specific values. Unlike discrete distributions, continuous distributions do not assign individual probabilities to each value but instead describe the distribution pattern across the range. Examples of continuous probability distributions include *normal (Gaussian) distribution, exponential distribution*, and *uniform distribution.*
   - *Gaussian distribution*: Gaussian distribution is extensively employed in statistics and probability theory. It is recognized for its bell-shaped curve and exhibits symmetry around its mean. It is often used to model real-world phenomena that tend to cluster around a central value with symmetrically decreasing probabilities as values move away from the mean. Many natural and social phenomena follow this distribution, such as heights and weights of individuals, errors in measurements, and IQ scores. Properties of Gaussian distribution, such as the central limit

theorem, make it a fundamental tool in statistical inference. It allows for the calculation of probabilities, confidence intervals, and hypothesis testing in various statistical analyses. To summarize, Gaussian distribution plays a vital role in statistical analysis and inference due to its mathematical properties and its prevalence in real-world data.

- *Exponential distribution*: This represents the time duration between events in a Poisson process. It is frequently employed to analyze the time taken for an event to occur, such as the waiting time for a customer in a queue or the interval between phone calls at a call center.
- *Uniform distribution*: This models outcomes where all values within a given interval are equally likely to occur. It is often used when there is no particular preference or bias toward any specific value within the range. Uniform distribution is often used in situations where there is an equal likelihood of an event or value occurring within a specified range. For example, it can be used to model the probability of selecting a random number between a and b, where each number has an equal chance of being chosen. Uniform distribution is also useful in simulation studies, generating random numbers, and as a baseline for comparing other probability distributions. In summary, uniform distribution is where all values within a defined interval have an equal probability of occurring. It is characterized by its lower and upper bounds, and its probability density is constant within the interval. Uniform distribution is commonly used in various applications, such as random number generation and simulation studies.

Probability distributions provide several important features and properties, which help summarize the central tendency, spread, and shape of the data. These properties are derived from the specific form and parameters of the distribution. Statisticians and researchers often use probability distributions to model real-world phenomena, make predictions, and perform statistical inference. By fitting observed data to a suitable probability distribution, they can estimate parameters, assess the likelihood of certain events, calculate probabilities, and perform hypothesis tests. A probability distribution is a mathematical function that assigns probabilities to diverse results or values of a random variable. It provides a framework for analyzing uncertainties, making predictions, and drawing conclusions based on observed data. Probability distributions play a vital role in statistical analysis, modeling real-world phenomena, and facilitating informed decision-making in various fields.

1.8 VARIOUS SAMPLE STATISTICS

Sample statistics are numerical measures that summarize and describe the characteristics of a sample, which is a subset of a larger complete set [20]. They provide valuable insights into the data and help in making inferences about the complete set. Here are some commonly used sample statistics:

- ***Sample mean*** $(\bar{\mathbf{x}})$**:** Average of all values in the sample and provides an estimate of the complete set mean.

- ***Sample median*:** The median corresponds to the central value within a sorted set of data, effectively dividing the data into two equal halves where 50% of the values lie below and 50% lie above it. It is less sensitive to extreme values than the mean.
- ***Sample mode*:** Most frequently occurring value or values in the sample. It provides insights into the data's central tendency and is particularly useful for categorical or discrete data.
- ***Sample variance* (s^2):** The variance quantifies the extent of dispersion or spread exhibited.
- Sample standard deviation (s): $\sqrt{s^2}$
- ***Sample range*:** *Max* – *Min*. It serves as a straightforward measure of the spread or extent of variation present in the data.
- ***Sample quartiles*:** Quartiles partition the data into four equal parts, denoting the 25th, 50th (median), and 75th percentiles. They offer valuable insights into the distribution of the data, enabling a deeper understanding of its characteristics.

These are just a few examples of sample statistics that are commonly used in statistical analysis. Each statistic provides unique information about the sample data and helps in understanding its characteristics and making inferences about the complete set.

1.9 ESTIMATION STATISTICS

Estimation statistics refers to the branch of statistics that deals with estimating unknown complete set parameters based on sample data. It encompasses the utilization of sample statistics to draw conclusions about complete set parameters and to measure the uncertainty linked with these estimations.

There are two main types of estimation in statistics:

- **Point estimation:** This comprises the process of estimating a singular value, referred to as a point estimate, which serves as the most suitable approximation of an unknown complete set parameter. Common examples of point estimators include the sample mean, sample proportion, or sample variance. Point estimates provide a single value as an estimate, but they do not indicate the variability, or the margin of error associated with the estimate.
- **Interval estimation:** This entails the creation of a range, referred to as a confidence interval, which is projected to contain the complete set parameter with a specified level of confidence. It signifies the degree of uncertainty linked with the estimate and provides a measure of its precision. Confidence intervals are frequently constructed by considering the characteristics of the sampling distribution and the desired level of confidence.

The process of estimation involves selecting an appropriate estimation method, collecting a representative sample from the complete set, calculating the sample statistics, and then using these statistics to estimate the unknown complete set parameter.

The choice of estimation method depends on various factors, including the type of data, the research question, and the assumptions underlying the estimation technique.

Estimation statistics holds immense significance across various domains, including market research, social sciences, public health, and economics. It empowers researchers to make informed decisions and derive conclusions about complete set parameters, even with limited sample data. Nevertheless, it is vital to acknowledge that estimation necessitates assumptions, and the accuracy of the estimates relies on the quality and representativeness of the sample.

To summarize, estimation statistics encompasses the process of utilizing sample data to estimate complete set parameters. It encompasses both point estimation, where a single value serves as the estimate, and interval estimation, where a range of values indicates the associated uncertainty. Estimation plays a vital role in making inferences and drawing conclusions based on limited sample data, but it is important to consider the assumptions and limitations of the estimation method used.

1.10 SAMPLE PROPORTIONS

The sample proportion is a statistic that represents the proportion or percentage of individuals in a sample that possess a certain characteristic or exhibit a specific behavior. It is commonly used to estimate the complete set proportion. It is calculated by dividing the number of individuals in the sample who have the desired characteristic or behavior by the total sample size.

The sample proportion formula (1.8) is used to calculate the proportion or percentage of individuals in a sample that possess a certain characteristic or exhibit a specific behavior. It is represented by the symbol $\hat{p}$.

$$\hat{p} = \left(\frac{x}{n}\right) \tag{1.8}$$

- $\hat{p}$ represents the sample proportion.
- x represents the desired characteristics of individuals
- n is the total sample size.

For example, if finding the proportion of adults in a certain city who support a particular political candidate is a problem, then take an arbitrary trial of 500 adults from the complete set and find that 250 individuals in the sample express support for the candidate. The sample proportion of adults who support the candidate would be 250/500 = 0.5 or 50%.

The properties of the sample proportion help assess the precision and variability of the estimate. It is crucial to recognize that the sample proportion is subject to sampling variability, implying that different samples extracted from the same complete set will likely yield slightly different estimates.

To summarize, the sample proportion serves as a statistic representing the proportion or percentage of individuals in a sample possessing a specific characteristic or exhibiting a particular behavior. Calculation involves dividing the count of individuals with the desired characteristic by the total sample size. The sample proportion enables estimation of the complete set proportion and facilitates implications about a complete set based on trial data.

1.11 CHI-SQUARE STATISTICS (X^2)

The chi-square statistic is an important measure involved in statistics to assess the independence or association between categorical variables. It relies on the chi-square distribution that characterizes the distribution of the sum of squared standard normal deviates.

Calculation of the chi-square statistic involves comparing observed frequencies with expected frequencies. It determines whether the observed frequencies significantly deviate from the expected frequencies that would be anticipated if the variables were independent.

The formula to calculate the chi-square statistic depends on the specific type of analysis being performed. Here are two common scenarios:

1. **Goodness of Fit Test:** In this scenario, the chi-square statistic is used to test whether observed frequencies in a single categorical variable significantly differ from expected frequencies based on a specified distribution. The formula for the chi-square statistic in this case is:

$$\chi^2 = \Sigma\left[(\text{Observed Frequency} - \text{Expected Frequency})^2/\text{Expected Frequency}\right] \quad (1.9)$$

Where Σ represents the sum over all categories.

2. **Test of Independence:** In this scenario, the chi-square statistic is used to assess the association between two categorical variables. The formula for the chi-square statistic in this case is:

$$\chi^2 = \Sigma\left[\frac{(\text{Observed Frequency} - \text{Expected Frequency})^2}{\text{Expected Frequency}}\right] \quad (1.10)$$

where Σ represents the sum over all cells of a contingency table.

The chi-square (χ^2) statistic is a valuable tool for examining the independence of categorical variables. By assessing the degree of deviation, the chi-square statistic helps determine whether the observed differences are statistically significant. This test is widely utilized in social sciences, market research, biology, and other fields to evaluate relationships among categorical variables and gain insights into their dependence or independence.

For instance, consider the example of tossing 100 counts of a fair coin. Outcome assumed is 50 counts head and 50 counts tail. However, the actual results may deviate from this expectation, such as yielding 60 heads and 40 tails or 90 heads and 10 tails. The chi-square test allows us to measure how well the observed results align with the theoretical expectation of a fair coin. If the observed results significantly differ from the expected outcome, it suggests that the coin may not be fair.

In summary, the chi-square statistic serves as a valuable measure for testing the independence or association between categorical variables. Widely employed in statistical analysis, the chi-square test provides insights into the relationships and dependencies among categorical data.

1.12 CENTRAL LIMIT THEOREM

The central limit theorem (CLT) is a crucial concept in statistics that outlines the behavior of sample means or sums derived from any complete set, regardless of the underlying distribution. It asserts that as the sample size increases, the distribution of sample means or sums converges toward a normal distribution, irrespective of the shape of the complete set distribution.

Mathematically, the central limit theorem can be expressed as follows:

$$X_1, X_2 \ldots, X_n - \text{complete set random variables}$$

$$\mu - \text{mean}$$

$$\sigma - \text{standard deviation}$$

$$n - \text{sample size}$$

To illustrate, let's consider the heights of individuals in a complete set. The complete set distribution may not follow a normal distribution. However, if we gather random samples of heights and compute the means for each sample, the distribution of these sample means will exhibit a close approximation to a normal distribution.

The central limit theorem holds significant implications for statistical inference and hypothesis testing. Additionally, it forms the basis for constructing confidence intervals and conducting hypothesis tests.

The theorem is widely used in various statistical applications. Consider:

- In market research, the central limit theorem is applied to estimate the average rating of a product based on a sample of consumer ratings.
- In quality control, it is used to assess the variability in product measurements and determine control limits.

The CLT is a significant notion in statistics because it permits us to make reliable implications about complete set parameters based on trial data. It provides a bridge between the characteristics of a complete set and the statistical properties of samples, enabling the use of powerful techniques such as hypothesis testing and confidence interval estimation.

In summary, the CLT is a fundamental concept in statistics, providing the basis for making inferences about complete set parameters. The theorem is widely used in various statistical applications to analyze data and draw meaningful conclusions.

1.13 SUMMARY

In conclusion, statistics plays a crucial role in various fields and disciplines by providing tools and techniques to analyze and interpret data. It allows researchers and decision-makers to make informed decisions based on evidence and quantifiable information. This chapter provides a comprehensive overview and understanding of statistics.

The importance of statistics lies in its ability to uncover patterns, relationships, and trends in data, enabling us to make predictions, test hypotheses, and draw meaningful insights. Whether it's estimating complete set parameters, assessing the significance of differences, or making forecasts, statistics provides a rigorous framework for analyzing data and making evidence-based decisions. It helps in understanding the variability and uncertainty inherent in data, and provides methods to quantify and account for these uncertainties.

Overall, statistics serves as a powerful tool for researchers, businesses, policymakers, and individuals alike. It allows us to explore and make sense of complex information, make reliable inferences about complete sets, and guide decision-making in a wide range of domains. As the world continues to generate vast amounts of data, the role of statistics will only grow in importance, providing invaluable insights and contributing to advancements in knowledge and understanding.

REFERENCES

1. Online Statistics Education. *A Multimedia Course of Study* (http://onlinestatbook.com/). Project Leader: David M. Lane, Rice University.
2. He, Xuming, and Xihong Lin. "Challenges and Opportunities in Statistics and Data Science: Ten Research Areas." *Harvard Data Science Review* 2.3 (2020): 1–8.
3. Gupta, S. C., and V.K. Kapoor. *Fundamentals of Mathematical Statistics*. Sultan Chand & Sons, 2020.
4. Lauritzen, Steffen. *Fundamentals of Mathematical Statistics*. CRC Press, 2023.
5. Grami, Ali. *Probability, Random Variables, Statistics, and Random Processes: Fundamentals & Applications*. John Wiley & Sons, 2019.
6. Lock, Robin H., et al. *Statistics: Unlocking the Power of Data*. John Wiley & Sons, 2020.
7. Johnson, Richard A., and Gouri K. Bhattacharyya. *Statistics: Principles and Methods*. John Wiley & Sons, 2019.
8. Coolidge, Frederick L. *Statistics: A Gentle Introduction*. Sage Publications, 2020.
9. Adams, Kathrynn A., and Eva K. McGuire. *Research Methods, Statistics, and Applications*. Sage Publications, 2022.
10. Riemann, B.L., and Lininger, M.R. "Principles of Statistics: What the Sports Medicine Professional Needs to Know." *Clinics in Sports Medicine* 37.3 (2018 Jul): 375–386. doi: 10.1016/j.csm.2018.03.004. PMID: 29903380.
11. Beyer, William H. *Handbook of Tables for Probability and Statistics*. CRC Press, 2019.
12. Kaliyadan, Feroze, and Vinay Kulkarni. "Types of Variables, Descriptive Statistics, and Sample Size." *Indian Dermatology Online Journal* 10.1 (2019): 82.
13. Bakker, Caitlin. "An Introduction to Statistics for Librarians (Part Two): Frequency Distributions and Measures of Central Tendency." *Hypothesis: Research Journal for Health Information Professionals* 35.1 (2023): 1–5.

14. Alnemrawy, Ziad. "Levels of Statistical Thinking in the Measures of Central Tendency (Mean, Median, and Mode) among Eighth-Graders in Jordan." *Journal of Physics Conference Series* 80.80 (2020): 1215–1245.
15. Amrhein, Valentin, David Trafimow, and Sander Greenland. "Inferential Statistics as Descriptive Statistics: There Is No Replication Crisis If We Don't Expect Replication." *The American Statistician* 73.sup1 (2019): 262–270.
16. Campbell, Michael J., and Richard M. Jacques. *Statistics at Square Two*. John Wiley & Sons, 2023.
17. Ross, Sheldon M. *Introduction to Probability and Statistics for Engineers and Scientists*. Academic Press, 2020.
18. Baron, Michael. *Probability and Statistics for Computer Scientists*. CRC Press, 2019.
19. Kounev, Samuel, et al. "Review of Basic Probability and Statistics." *Systems Benchmarking: For Scientists and Engineers* 1 (2020): 23–44.
20. Lohr, Sharon L. *Sampling: Design and Analysis*. CRC Press, 2021.

2 Fundamentals of Quantum Machines

Soham S. Bhoir and Harshal H. Dave
KJ Somaiya College of Engineering, India

Devi Priya Rangasamy
KPR Institute of Engineering and Technology, India

2.1 INTRODUCTION TO QUANTUM MACHINES

Quantum machines represent a groundbreaking paradigm shift in the realm of computation, ushering in an era of limitless possibilities. Harnessing the intricate principles of quantum mechanics, they promise an unparalleled leap in computational prowess. Unlike their classical counterparts, which rely on binary classical bits, quantum machines harness the enigmatic power of qubits: quantum bits endowed with extraordinary characteristics. Qubits can exist in a superposition of states, allowing them to represent multiple values simultaneously, a phenomenon inconceivable in classical computing. Furthermore, the phenomenon of entanglement, where qubits become interconnected and instantaneously influence each other, unveils a whole new dimension of computing capabilities. This convergence of superposition and entanglement forms the cornerstone for quantum algorithms, capable of tackling intricate problems with an exponential speedup, unlocking solutions that once appeared insurmountable. The dawn of quantum computation promises to redefine the boundaries of what we can achieve in fields ranging from cryptography to material science, revolutionizing industries and fueling scientific breakthroughs.

The motivation behind the development of quantum machines lies in the quest for tackling computational challenges that surpass the capabilities of classical computers. Traditional computers encounter difficulties when confronted with complex optimization problems, simulation of quantum systems, and prime factorization, among others. Quantum machines, with their ability to manipulate vast amounts of information simultaneously, hold immense potential to revolutionize these domains and unlock new avenues of scientific and technological exploration.

2.1.1 Comparison with Classical Machines

A profound comprehension of the distinctions between quantum and classical machines sets the stage for discerning the unique advantages offered by the former. Classical machines operate on classical bits, which occupy a binary state of either 0 or 1. In stark

DOI: 10.1201/9781003429654-3

contrast, qubits have the extraordinary ability to exist in a superposition of states, allowing them to embody multiple possibilities concurrently. This inherent parallelism grants quantum machines an exponential surge in computational capacity, enabling them to solve problems of immense complexity that would baffle classical systems [1].

Another distinctive hallmark that elevates quantum machines above classical counterparts is the fascinating phenomenon of entanglement. Within the enchanting realm of quantum physics, qubits, the fundamental units of quantum information, forge intricate correlations that transcend spatial boundaries. This mesmerizing interconnection imbues quantum machines with a unique capability to execute computations in a profoundly synchronized and interwoven fashion. It's as if these qubits communicate instantaneously, defying the constraints of space and time. This remarkable attribute of quantum systems opens the door to tantalizing prospects in various domains. From bolstering the security of cryptographic protocols, to revolutionizing ultra-secure communication systems, and even enhancing optimization algorithms, entanglement becomes a powerful tool propelling us towards unprecedented advancements in science and technology [2].

As we embark on this enlightening journey, note that the upcoming sections will delve into the intricacies of quantum computing with a meticulous focus on foundational concepts, theoretical underpinnings, and practical applications.

2.2 QUANTUM MECHANICS FOR QUANTUM COMPUTING

Quantum mechanics, the foundational theory of modern physics, revolutionized our understanding of the microscopic world, providing a mathematical framework to describe the behavior of particles at the quantum level. This remarkable theory, developed in the early 20th century, forms the cornerstone of quantum computing, unlocking the potential for exponentially powerful computation through the exploitation of quantum phenomena.

The "father of quantum mechanics," Werner Heisenberg, played a pivotal role in shaping this field. In 1925, Heisenberg, along with Max Born and Pascual Jordan, formulated the revolutionary mathematical formalism of quantum mechanics, known as matrix mechanics. Heisenberg's seminal work introduced the uncertainty principle, which states that certain pairs of physical properties, such as position and momentum, cannot be simultaneously measured with arbitrary precision. This principle challenged classical notions of determinism and laid the foundation for the probabilistic nature of quantum phenomena.

Heisenberg's uncertainty principle, coupled with the groundbreaking contributions of other luminaries such as Erwin Schrödinger, Paul Dirac, and Niels Bohr, led to the development of wave mechanics, an alternative formulation of quantum mechanics based on the concept of wavefunctions. Schrödinger's wave equation, proposed in 1926, provided a means to describe the time evolution of quantum systems and to calculate their wavefunctions.

2.2.1 QUANTUM STATES AND WAVEFUNCTIONS

In the realm of quantum mechanics, quantum states serve as the fundamental descriptions of quantum systems. They encapsulate the properties and potential outcomes of these systems, laying the groundwork for subsequent calculations and predictions. To comprehend the intricacies of quantum states, we turn to wavefunctions, the mathematical representations that allow us to explore the quantum realm. A wavefunction, denoted as $|\psi\rangle$, captures the complete information about the state of a quantum system. It is a complex-valued function defined within the framework of Hilbert space, a mathematical construct that provides a rigorous foundation for quantum mechanics. In this formalism, quantum states are represented by vectors in Hilbert space, and their evolution over time is governed by the principles of unitary transformations. Mathematically, a wavefunction $|\psi\rangle$ can be expressed as a linear combination of basis states, represented as $|\psi\rangle = \Sigma_i \alpha_i |\Phi_i\rangle$, where α_i are complex coefficients and $|\Phi_i\rangle$ are the orthonormal basis states. These coefficients, known as probability amplitudes, determine the probabilities of obtaining various measurement outcomes upon measurement [3].

The normalization condition ensures that the probabilities of all possible measurement outcomes sum up to unity. It requires that the integral of the absolute square of the wavefunction over the entire space is equal to 1:

$$\int |\psi(x)|^2 \, dx = 1 \tag{2.1}$$

This condition guarantees that the total probability of finding the quantum system in any possible state is conserved [4].

For instance, let's consider a particle confined to a one-dimensional box of length L. The wave function representing the particle's state can be expressed as:

$$\psi(x) = \sqrt{\frac{2}{L}} \sin\left(\frac{n\pi x}{L}\right) \tag{2.2}$$

where n is an integer representing the energy level of the particle. By normalizing this wavefunction, we ensure that the probability of finding the particle within the box is 1.

The concept of superposition, a hallmark of quantum mechanics, is further examined to deepen our understanding. Quantum systems are allowed to exist in a linear combination of multiple states simultaneously, which is mathematically represented as:

$$|\psi\rangle = \sum_i \alpha_i |\Phi_i\rangle \tag{2.3}$$

where α_i are complex coefficients and $|\Phi_i\rangle$ are the orthonormal basis states. This property of superposition gives rise to the famous thought experiment of Schrödinger's cat, where a cat can exist in a superposition of being both alive and dead until an observation is made [4, 5].

The act of measurement in quantum mechanics is a process that collapses the wavefunction onto one of its possible eigenstates, corresponding to a definite measurement outcome. The probability of obtaining a particular measurement outcome is given by the Born rule, which states that the probability of measuring the system in the state $|\Phi_i\rangle$ is equal to the squared modulus of the coefficient α_i:

$$P(\Phi_i) = |\alpha_i|^2 . \tag{2.4}$$

This probabilistic nature of quantum mechanics is a departure from classical physics, where properties of systems are determined with certainty. In quantum mechanics, measurements only yield probabilistic results, and it is the repeated measurements on an ensemble of identically prepared systems that reveal the statistical behavior. Quantum states also exhibit another peculiar characteristic called entanglement. When two or more quantum systems become entangled, their individual wavefunctions can no longer be described independently. Instead, the entangled system is described by a joint wavefunction that accounts for the correlations between the systems. This entanglement can result in non-local correlations, famously described by Einstein, Podolsky, and Rosen in their EPR paradox. Furthermore, the evolution of quantum states over time is described by the Schrödinger equation, which is a central equation in quantum mechanics. The Schrödinger equation dictates how the wavefunction of a quantum system changes in response to its Hamiltonian, which represents the total energy of the system [3–5].

In its time-independent form, the Schrödinger equation is given by:

$$\hat{H}\left|\psi\right\rangle = E\left|\psi\right\rangle \tag{2.5}$$

where $\hat{H}$ is the Hamiltonian operator and E is the energy eigenvalue associated with the state $|\psi\rangle$. Solving the Schrödinger equation allows us to determine the allowed energy levels and corresponding wavefunctions of a quantum system.

The concept of observables in quantum mechanics is intimately connected to quantum states. Observables are physical quantities that can be measured, such as position, momentum, energy, and spin. In quantum mechanics, observables are represented by corresponding operators, which act on the wavefunction to yield measurement outcomes. The eigenvalues of the operator represent the possible measurement results, while the eigenstates correspond to the states in which the measurements are certain. For example, the position of a particle is represented by the position operator $\hat{x}$, and its momentum is represented by the momentum operator $\hat{p}$. When acting on the wavefunction, these operators yield the position and momentum measurement outcomes, respectively. The probabilities of obtaining specific measurement results are given by the squared modulus of the corresponding coefficients in the wavefunction expansion. It is important to note that the act of measurement in quantum mechanics can induce a collapse of the wavefunction onto an eigenstate of the measured observable. This collapse is a non-deterministic process, and the outcome of a measurement is inherently uncertain. This characteristic is often referred to as the measurement problem in quantum mechanics and has been a subject of philosophical debates and interpretations.

The concept of quantum states and wavefunctions is not limited to single particles. It extends to systems with multiple particles, such as atoms, molecules, and even macroscopic objects. The wavefunction of a multi-particle system is described by a multi-variable function that includes the positions or other relevant variables of all the particles involved. It also forms the foundation for various quantum phenomena and applications. From the principles of superposition and entanglement, quantum computing, quantum communication, and quantum cryptography have emerged as promising fields. Quantum states and their manipulation allow for the development of quantum algorithms that can solve certain problems more efficiently than classical algorithms.

2.3 QUANTUM BITS AND STATES

Quantum computing harnesses the principles of quantum mechanics to process and manipulate information in ways that surpass the capabilities of classical computing. At the core of quantum computing are quantum bits, or qubits, which serve as the fundamental units of information [6]. In this section, we explore the concept of qubits, their representation, and their unique properties. Additionally, we delve into the notion of quantum states and their mathematical descriptions that form the basis for quantum computation.

2.3.1 QUBITS: THE BUILDING BLOCKS OF QUANTUM COMPUTING

In classical computing, the fundamental unit of information is the classical bit, which can exist in one of two states: 0 or 1. Quantum computing, on the other hand, introduces qubits as the fundamental units of information. Qubits are quantum systems that can exist in a superposition of both 0 and 1 states simultaneously, allowing for a much richer and more powerful representation of information. The physical realization of a qubit can vary depending on the underlying technology. For example, qubits can be implemented using trapped ions, superconducting circuits, or quantum dots. Regardless of the physical implementation, qubits must exhibit two key properties: superposition and entanglement [7].

2.3.2 SUPERPOSITION

Superposition is a fundamental property of qubits that allows them to exist in a coherent combination of multiple states simultaneously. In other words, a qubit can be in a state that represents both 0 and 1 at the same time, with the specific probabilities of each state determined by the coefficients in its quantum state vector.

Mathematically, a qubit in superposition can be represented as:

$$|\psi\rangle = \alpha|0\rangle + \beta|1\rangle, \tag{2.6}$$

where α and β are complex probability amplitudes, and $|0\rangle$ and $|1\rangle$ represent the basis states of the qubit. The coefficients α and β must satisfy the normalization condition $|\alpha|^2 + |\beta|^2 = 1$ to ensure that the total probability of measuring the qubit in either state is conserved.

Superposition enables qubits to simultaneously explore multiple computational paths and perform computations in parallel, providing the potential for significant speedup in certain algorithms [7].

2.3.3 ENTANGLEMENT

Entanglement is another crucial property of qubits that allows for strong correlations between multiple qubits, even when they are physically separated. When qubits are entangled, their quantum states become interdependent, and the measurement outcome of one qubit can instantaneously influence the state of the other, regardless of the distance between them. This phenomenon, often referred to as "spooky action at a distance," defies classical intuition but is a fundamental feature of quantum mechanics. Entangled qubits are described by a joint quantum state that cannot be decomposed into individual qubit states. The state of a two-qubit entangled system can be expressed as:

$$|\psi\rangle = \alpha|00\rangle + \beta|01\rangle + \gamma|10\rangle + \delta|11\rangle, \tag{2.7}$$

where α, β, γ, and δ are complex probability amplitudes. The entanglement between qubits allows for the representation of highly correlated information and enables quantum algorithms such as quantum teleportation and quantum error correction. Entanglement is a valuable resource in quantum computing as it enables the exploitation of parallelism and enables novel computational capabilities that surpass classical systems [7].

2.3.4 Quantum States and their Mathematical Description

Quantum states provide a mathematical description of the quantum system and encode the information needed for quantum computations. Quantum states are represented by vectors in a complex vector space known as Hilbert space. The evolution of quantum states over time is governed by the principles of unitary transformations.

Mathematically, a quantum state can be represented as:

$$|\psi\rangle = \sum_i \alpha_i |\phi_i\rangle, \tag{2.8}$$

where α_i are complex probability amplitudes, and $|\phi_i\rangle$ are the orthonormal basis states that span the Hilbert space. The coefficients α_i determine the probabilities of obtaining various measurement outcomes when the quantum system is measured [8, 9].

The normalization condition ensures that the total probability of all possible measurement outcomes sums to unity. It requires that the inner product of the quantum state with itself is equal to 1:

$$\langle\psi|\psi\rangle = \sum_i |\alpha_i|^2 = 1. \tag{2.9}$$

This normalization condition guarantees the conservation of probability and ensures that the quantum system is in a valid state [7, 8].

The mathematical description of quantum states allows for the manipulation and transformation of the state vector using various quantum operations. Unitary operators, represented by matrices, perform transformations on the quantum state without changing the normalization or the inner product. These unitary transformations correspond to the evolution of the quantum system under different operations such as quantum gates and measurements. The evolution of quantum states over time is described by the Schrödinger equation, which is a fundamental equation in quantum mechanics. It governs the dynamics of the quantum system and determines how the quantum state changes in response to the system's Hamiltonian. In quantum computing, quantum algorithms manipulate quantum states through a sequence of quantum operations to perform specific computational tasks efficiently. By harnessing the properties of superposition and entanglement, quantum algorithms can solve certain problems exponentially faster than classical algorithms.

2.4 UNDERSTANDING QUANTUM COMPUTING

2.4.1 Quantum Gates and Operations

In classical computing, logic gates form the building blocks of digital circuits and enable the manipulation of classical bits. Similarly, in quantum computing, quantum gates serve as the fundamental operations for manipulating qubits. Quantum gates are unitary operators that transform the quantum state of one or more qubits. The choice and arrangement of quantum gates determine the computational operations performed on the quantum state. These gates can be combined to create quantum circuits that implement specific algorithms or perform desired computations [6].

2.4.1.1 Single-Qubit Gates

Single-qubit gates act on individual qubits and allow for the manipulation of their quantum states. Some commonly used single-qubit gates include:

- **Pauli-*X* Gate (*X* gate):** The Pauli-*X* gate is analogous to the classical NOT gate and flips the state of a qubit from $|0\rangle$ to $|1\rangle$ and vice versa. Mathematically, it can be represented as:

$$X = \begin{bmatrix} 0 & 1 \\ 1 & 0 \end{bmatrix}. \tag{2.10}$$

- **Pauli-*Y* Gate (*Y* gate):** The Pauli-*Y* gate is another single-qubit gate that introduces a phase shift between the basis states. It rotates the state of a qubit around the *y*-axis of the Bloch sphere. Mathematically, it can be represented as:

$$Y = \begin{bmatrix} 0 & -i \\ i & 0 \end{bmatrix}. \tag{2.11}$$

- **Pauli-*Z* Gate (*Z* gate):** The Pauli-*Z* gate applies a phase flip to the qubit state, leaving $|0\rangle$ unchanged and flipping the sign of $|1\rangle$. Mathematically, it can be represented as:

$$Z = \begin{bmatrix} 1 & 0 \\ 0 & -1 \end{bmatrix}. \tag{2.12}$$

- **Hadamard Gate (*H* gate):** The Hadamard gate is a versatile gate that creates superposition by transforming $|0\rangle$ to an equal superposition of $|0\rangle$ and $|1\rangle$. It also introduces a phase shift, rotating the state around the x-axis of the Bloch sphere. Mathematically, it can be represented as:

$$H = \frac{1}{\sqrt{2}}\begin{bmatrix} 1 & 1 \\ 1 & -1 \end{bmatrix}. \tag{2.13}$$

These single-qubit gates can be combined and applied sequentially or in parallel to perform complex operations on a quantum state [10, 11].

2.4.1.2 Multi-Qubit Gates

Multi-qubit gates allow for the interaction and entanglement of multiple qubits, enabling the implementation of quantum algorithms. Some commonly used multi-qubit gates include:

- **Controlled-NOT Gate (CNOT gate):** The CNOT gate, also known as the controlled-*X* gate, is a two-qubit gate that performs an *X* gate operation on the target qubit (the second qubit) conditioned on the state of the control qubit (the first qubit). It can be represented as:

$$\text{CNOT} = \begin{bmatrix} 1 & 0 & 0 & 0 \\ 0 & 1 & 0 & 0 \\ 0 & 0 & 0 & 1 \\ 0 & 0 & 1 & 0 \end{bmatrix}. \tag{2.14}$$

  The CNOT gate is crucial for creating entanglement between qubits and performing quantum computations.
- **Toffoli Gate (CCNOT gate):** The Toffoli gate is a three-qubit gate that acts as a controlled-controlled-NOT gate. It performs an *X* gate operation on the target qubit (the third qubit) conditioned on the states of the two control qubits (the first and second qubits). It can be represented as:

$$\text{CCNOT} = \begin{bmatrix} 1 & 0 & 0 & 0 & 0 & 0 & 0 & 0 \\ 0 & 1 & 0 & 0 & 0 & 0 & 0 & 0 \\ 0 & 0 & 1 & 0 & 0 & 0 & 0 & 0 \\ 0 & 0 & 0 & 1 & 0 & 0 & 0 & 0 \\ 0 & 0 & 0 & 0 & 1 & 0 & 0 & 0 \\ 0 & 0 & 0 & 0 & 0 & 1 & 0 & 0 \\ 0 & 0 & 0 & 0 & 0 & 0 & 0 & 1 \\ 0 & 0 & 0 & 0 & 0 & 0 & 1 & 0 \end{bmatrix}. \tag{2.15}$$

  The Toffoli gate is a universal gate, meaning that any quantum computation can be expressed using a combination of Toffoli gates and single-qubit gates.

- **Swap Gate:** The swap gate exchanges the states of two qubits, effectively swapping their quantum information. It can be represented as:

$$\text{SWAP} = \begin{bmatrix} 1 & 0 & 0 & 0 \\ 0 & 0 & 1 & 0 \\ 0 & 1 & 0 & 0 \\ 0 & 0 & 0 & 1 \end{bmatrix}. \quad (2.16)$$

The swap gate is particularly useful in quantum algorithms and quantum error correction.

These multi-qubit gates, combined with single-qubit gates, provide the necessary tools to perform various computations and transformations on quantum states [6, 10, 11].

2.4.1.3 Quantum Circuit Examples

To better understand the use of quantum gates, let's consider a simple example of a quantum circuit. Suppose we have two qubits, initially prepared in the state $|00\rangle$. We want to apply a Hadamard gate to the first qubit, followed by a CNOT gate with the first qubit as the control and the second qubit as the target. The resulting quantum circuit can be represented as:

q_0 —[H]—●—
q_1 ————⊕—

Analyzing the evolution of the quantum state through this circuit:

1. Initially, the two qubits are prepared in the state $|00\rangle$, represented as:

$$|00\rangle = \begin{bmatrix} 1 \\ 0 \\ 0 \\ 0 \end{bmatrix} \quad (2.17)$$

2. The Hadamard gate (H gate) is applied to the first qubit. This transforms the state as follows:

$$\frac{1}{\sqrt{2}}\left(|0\rangle + |1\rangle\right) \otimes |0\rangle = \frac{1}{\sqrt{2}}\left(|00\rangle + |10\rangle\right). \quad (2.18)$$

3. The controlled-NOT gate (CNOT gate) is applied with the first qubit as the control and the second qubit as the target. This gate flips the state of the target qubit (second qubit) if the control qubit (first qubit) is in state $|1\rangle$. The resulting state is:

$$\frac{1}{\sqrt{2}}\left(|00\rangle + |11\rangle\right). \quad (2.19)$$

The quantum state after the application of the Hadamard gate and the CNOT gate represents an entangled state, where both qubits are correlated and cannot be described independently [12].

This simple example demonstrates the application of quantum gates in manipulating quantum states and creating entanglement between qubits. Quantum circuits, composed of various gates, enable the execution of quantum algorithms and computations.

2.4.2 Quantum Applications

Quantum computing holds the potential to revolutionize various fields by solving problems that are computationally intractable for classical computers. In this section, we explore some notable quantum applications and their underlying principles.

2.4.2.1 Quantum Simulation

Quantum simulation aims to utilize quantum computers to simulate and study quantum systems that are difficult to analyze classically. By leveraging the inherent quantum properties, such as superposition and entanglement, quantum simulations can provide insights into the behavior of complex molecules, materials, and physical phenomena. For example, simulating the electronic structure of molecules plays a crucial role in drug discovery and materials science. The quantum simulation approach allows for a more accurate and efficient representation of molecular systems, enabling researchers to explore potential drug candidates and optimize material properties [13, 14].

2.4.2.2 Optimization

Optimization problems are prevalent in various fields, ranging from logistics and finance to supply chain management and machine learning. Quantum computers have the potential to outperform classical algorithms in solving optimization problems by leveraging quantum algorithms such as the quantum approximate optimization algorithm (QAOA) and the quantum annealing algorithm (QAA). Quantum optimization algorithms exploit quantum effects to efficiently search for optimal solutions in large search spaces. They leverage concepts from quantum mechanics, such as quantum superposition and quantum interference, to explore multiple potential solutions simultaneously and converge on the optimal solution more effectively than classical methods [15].

2.4.2.3 Cryptography

Cryptography is crucial for ensuring secure communication and data protection. Classical cryptographic systems, such as the widely used RSA encryption, rely on the difficulty of factoring large numbers and the hardness of certain mathematical problems for their security. However, Shor's algorithm poses a significant threat to these classical cryptographic systems. Quantum computing offers the potential to break classical cryptographic systems by leveraging the speedup provided by quantum algorithms. Quantum-resistant cryptography, also known as post-quantum cryptography, aims to develop encryption schemes and cryptographic protocols that are resistant to attacks from quantum computers. The exploration and development of post-quantum cryptographic algorithms are essential to ensure secure communication in the post-quantum era [16].

These are just a few examples of the potential applications of quantum computing. As the field continues to advance, we can expect quantum computers to impact various domains, revolutionizing computational power and unlocking new possibilities for scientific exploration and technological advancements.

2.4.2.4 Machine Learning

Machine learning is a rapidly growing field that relies on computational power for training complex models and making predictions. Quantum computing has the potential to enhance machine learning algorithms and enable more efficient processing of large datasets. Quantum machine learning algorithms, such as quantum support vector machines and quantum neural networks, leverage the power of quantum computers to process and analyze data in quantum states. These algorithms can potentially provide exponential speedups for certain machine learning tasks, such as pattern recognition and optimization of model parameters. Quantum machine learning also explores the use of quantum-inspired algorithms on classical computers. These algorithms, inspired by the principles of quantum mechanics, aim to exploit certain quantum-like effects to improve the performance of classical machine learning algorithms [17].

2.4.2.5 Chemistry and Materials Science

Quantum computing offers promising opportunities in the fields of chemistry and materials science, where understanding the behavior of molecules and materials at the quantum level is crucial. Quantum computers can efficiently simulate the electronic structure of molecules, allowing for accurate predictions of chemical reactions and properties. The simulation of quantum systems can enable the discovery of new materials with tailored properties for various applications, such as energy storage, catalysis, and drug development. Quantum algorithms, such as the variational quantum eigensolver (VQE) and the quantum phase estimation (QPE) algorithm, provide avenues for solving complex quantum chemistry problems more efficiently than classical methods [18].

2.4.2.6 Financial Modeling and Portfolio Optimization

Financial modeling and portfolio optimization are essential tasks in finance and investment management. Quantum computing can offer advantages in analyzing and optimizing investment portfolios, risk management, and option pricing. Quantum algorithms, such as the quantum amplitude estimation (QAE) algorithm and quantum Monte Carlo methods, can improve the accuracy and speed of financial simulations and optimization processes. These algorithms leverage the power of quantum computing to efficiently explore a large number of possible investment scenarios, enabling more informed decision-making and portfolio management [19].

2.4.2.7 Transportation and Logistics

Transportation and logistics involve complex optimization problems, such as route planning, scheduling, and resource allocation. Quantum computing can provide solutions to these problems by enabling faster and more efficient search algorithms. Quantum algorithms, such as the quantum approximate optimization (QAOA) algorithm and the quantum traveling salesman problem (QTSP) algorithm, offer the potential for finding optimal or near-optimal solutions in a shorter time than classical methods. This can lead to improved route planning, cost reduction, and more efficient resource allocation in transportation and logistics operations. These are just a few examples of the potential applications of quantum computing across various fields. As the field of quantum computing continues to advance, we can expect to see more innovative applications emerge, transforming industries and driving new discoveries and breakthroughs [15].

2.5 QUANTUM PROTOCOLS AND QUANTUM ALGORITHMS

Quantum computing has revolutionized the field of information processing by offering unprecedented computational power and the ability to solve complex problems more efficiently than classical computers. In this section, we explore various quantum protocols and algorithms that harness the unique properties of quantum systems to achieve remarkable computational feats.

2.5.1 Quantum Protocols

Quantum protocols are a collection of techniques and methodologies that leverage the unique properties of quantum systems to achieve remarkable feats in various domains such as communication, cryptography, and simulation. These protocols exploit phenomena such as superposition, entanglement, and interference, enabling advancements that surpass the limitations of classical systems. In this section, we examine some prominent quantum protocols and their applications.

2.5.1.1 Quantum Communication Protocols

Quantum communication protocols harness the power of quantum systems to facilitate secure and efficient information exchange. They employ the principles of quantum mechanics to achieve secure key distribution and establish reliable communication channels:

1. Quantum Key Distribution (QKD)
   Quantum key distribution (QKD) protocols play a crucial role in secure communication by exploiting the fundamental principles of quantum mechanics. In a typical QKD scenario, Alice and Bob want to establish a shared secret key, while being cautious of any potential eavesdropper, Rudy. The security of QKD protocols is based on the laws of quantum mechanics, ensuring that any attempt by Rudy to intercept the quantum signals will be detected.

   The BB84 protocol, proposed by Bennett and Brassard in 1984, is a widely studied QKD protocol. It involves the transmission of qubits encoded in two complementary bases, typically represented as the computational basis ($|0\rangle$ and $|1\rangle$) and the Hadamard basis ($|+\rangle$ and $|-\rangle$). Alice randomly chooses one of the two bases and encodes her bit values accordingly. Bob also randomly chooses a basis for measurement. After transmission, Alice and Bob publicly announce their chosen bases and discard the measurements made in different bases. Through classical communication, they compare a subset of their remaining measurements to detect any discrepancies, indicative of eavesdropping attempts. The remaining correlated measurements are then used to generate a secure shared key [1].

2. Quantum Secure Direct Communication (QSDC)

   Quantum secure direct communication (QSDC) protocols allow secure communication between parties without the need for an established shared key. In a typical QSDC scenario, Alice wants to send a secret message directly to Bob, with Rudy unable to obtain any information about the message. Quantum entanglement plays a pivotal role in QSDC protocols.

   The Ping-Pong protocol, proposed by Deng and Long in 2003, is an example of a QSDC protocol. It utilizes an entangled state shared by Alice and Bob, such as a Bell state $\left(|\Phi^+\rangle = \frac{1}{\sqrt{2}}\left(|00\rangle + |11\rangle\right)\right)$. Alice encodes her secret message onto her particle of the entangled state and sends it to Bob. Bob performs a specific set of operations to retrieve the message, while Rudy, who lacks the entangled state, cannot obtain any information. The successful retrieval of the message by Bob ensures the secure communication between Alice and Bob [20].

2.5.1.2 Quantum Cryptography Protocols

Quantum cryptography protocols aim to enhance the security of classical cryptographic schemes by incorporating quantum principles. These protocols provide unbreakable encryption and secure communication channels by exploiting the laws of quantum mechanics.

1. Quantum Coin Flipping

   Quantum coin flipping protocols enable two parties, Alice and Bob, to establish a fair outcome for a coin flip, even in the presence of an untrusted third party, Rudy. The goal is to ensure that neither Alice nor Bob can manipulate the result, and the outcome is unbiased.

   The protocol, proposed by Blum in 1981, utilizes the properties of quantum systems to achieve fair coin flipping. It involves Alice and Bob sharing entangled particles, with each party randomly choosing a measurement basis. After performing measurements, Alice and Bob announce their measurement outcomes, and a fairness test is conducted to determine if the coin flip was indeed fair. If the fairness test is passed, the outcome of the coin flip is revealed [21].

2.5.1.3 Quantum Simulation Protocols

Quantum simulation protocols utilize quantum computers to simulate complex quantum systems that are challenging to study using classical computational methods. These protocols offer insights into the behavior of quantum matter, quantum chemistry, and condensed matter physics, paving the way for advances in various scientific disciplines. Quantum simulation protocols involve the careful design of quantum circuits and quantum algorithms tailored to the specific system under investigation. By encoding the properties of the target system into the quantum states and leveraging quantum operations, scientists can simulate the dynamics and properties of complex quantum systems. Examples include simulating the behavior of interacting particles, exploring the electronic structure of molecules, and studying quantum phase transitions [22].

2.5.2 Quantum Algorithms

Quantum algorithms form the backbone of quantum computing, harnessing the power of quantum systems to solve computational problems with remarkable efficiency. These algorithms take advantage of the unique properties of quantum bits (qubits), such as superposition and entanglement, to perform calculations in parallel and explore multiple possibilities simultaneously. In this section, we explore several notable quantum algorithms and their applications across different domains.

2.5.2.1 Deutsch-Jozsa Algorithm

The Deutsch-Jozsa algorithm, proposed by David Deutsch and Richard Jozsa in 1992, serves as a seminal example of a quantum algorithm that showcases the power of quantum computation. It aims to solve a specific type of problem known as the Deutsch-Jozsa problem, which involves determining whether a given function is constant or balanced [23].

2.5.2.2 The Problem

Consider a function f: $\{0, 1\}^n \rightarrow \{0, 1\}$ that takes as input an n-bit string and produces either a constant output (0 or 1) for all possible inputs or a balanced distribution of outputs, meaning half of the inputs yield 0 and the other half yield 1. The goal of the Deutsch-Jozsa algorithm is to determine the nature of the function (constant or balanced) with just a single query to the function.

2.5.2.3 The Classical Approach

In the classical setting, solving the Deutsch-Jozsa problem involves evaluating the function f for different input values to determine if it is constant or balanced. The classical solution requires multiple function evaluations to obtain a high probability of correctness [23].

Let's consider the example function f:

$$f(00) = 0, \tag{2.20}$$

$$f(01) = 1, \tag{2.21}$$

$$f(10) = 0, \tag{2.22}$$

$$f(11) = 1. \tag{2.23}$$

To determine if a function is constant or balanced classically, one can evaluate the function for different input values. The general form of the function is as follows:

$$f(00) = 0, \tag{2.24}$$

$$f(01) = 1, \tag{2.25}$$

$$f(10) = 0, \tag{2.26}$$

$$f(11) = 1. \tag{2.27}$$

From these evaluations, we can observe that the function produces both 0 and 1 outputs, indicating that it is balanced. However, in general, we would need to evaluate the function for $2^{(n-1)} + 1$ input values to determine if it is constant or balanced.

The classical approach to solving the Deutsch-Jozsa problem involves sequentially evaluating the function for different input values, as classical bits can only exist in states 0 or 1. This sequential evaluation limits the efficiency of the classical solution.

2.5.2.4 The Quantum Approach

The Deutsch-Jozsa algorithm exploits the principles of superposition and interference in quantum computation to solve the problem with a single query to the function [23].

2.5.2.5 The Quantum Circuit

The algorithm employs a quantum circuit that consists of the following steps:

Step 1: Initialization
Create an n-qubit quantum register in the state $|0^n\rangle$, where $|0^n\rangle$ represents the all-zero state.

Step 2: Superposition
Apply Hadamard gates (H) to all qubits to create an equal superposition of all possible input states:

$$|\psi_1\rangle = \frac{1}{\sqrt{2^n}} \sum_x |x\rangle, \tag{2.28}$$

where $|x\rangle$ represents an n-bit binary string.

Step 3: Function Evaluation
Apply an oracle transformation U_f to the quantum register, which encodes the function f as a unitary operator. This oracle acts on the input and output qubits as follows:

$$U_f |x, y\rangle = |x, y \oplus f(x)\rangle, \tag{2.29}$$

where $\oplus$ denotes bitwise addition modulo 2, and y is an auxiliary qubit initially in the state $|1\rangle$.

Step 4: Interference

Apply Hadamard gates to the input qubits once again:

$$|\psi_2\rangle = \frac{1}{\sqrt{2^n}} \sum_x (-1)^{f(x)} |x\rangle. \tag{2.30}$$

Step 5: Measurement

Measure the input qubits. If the measurement yields the all-zero state, the function is constant. Otherwise, it is balanced.

2.5.2.6 The Algorithm Execution

To understand the working of the Deutsch-Jozsa algorithm, let's consider a simple example. Suppose we have a 2-bit function *f*, defined as follows:

$$f(00) = 0, \tag{2.31}$$

$$f(01) = 1, \tag{2.32}$$

$$f(10) = 0, \tag{2.33}$$

$$f(11) = 1. \tag{2.34}$$

2.5.2.7 Initialization

We begin by preparing a quantum register in the state $|00\rangle$.

2.5.2.8 Superposition

Applyi Hadamard gates to the input qubits, to create a superposition:

$$|\psi_1\rangle = \frac{1}{2}\left(|00\rangle + |01\rangle + |10\rangle + |11\rangle\right) \tag{2.35}$$

2.5.2.9 Function Evaluation

Next, we encode the function f into the quantum circuit. The oracle transformation U_f acts as follows:

$$U_f |x, y\rangle = |x, y \oplus f(x)\rangle. \tag{2.36}$$

In our example, the oracle transforms the state as:

$$\begin{aligned} U_f |\psi_1\rangle &= \frac{1}{2}\left(|00 \oplus 0\rangle + |01 \oplus 1\rangle + |10 \oplus 0\rangle + |11 \oplus 1\rangle\right) \\ &= \frac{1}{2}\left(|00\rangle + |01\rangle + |10\rangle - |11\rangle\right) \end{aligned} \tag{2.37}$$

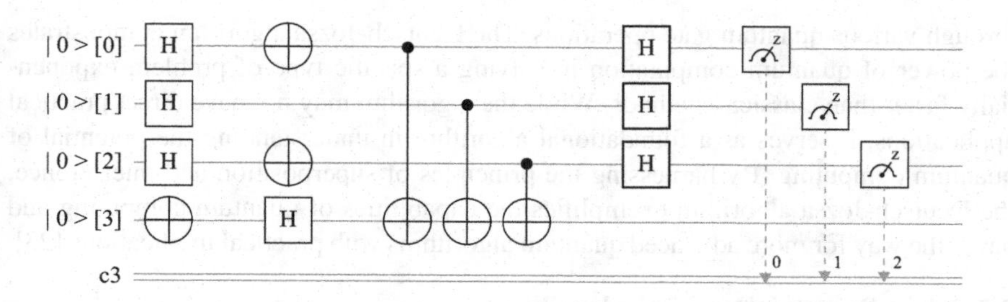

FIGURE 2.1 Circuit visualization of the Deutsch-Jozsa algorithm.

2.5.2.10 Interference

Applying Hadamard gates to the input qubits once again, we obtain:

$$|\psi_2\rangle = \frac{1}{2}\Big(\big(|00\rangle + |01\rangle + |10\rangle - |11\rangle\big) + \big(|00\rangle - |01\rangle + |10\rangle + |11\rangle\big) \\ + \big(|00\rangle + |01\rangle - |10\rangle - |11\rangle\big) - \big(|00\rangle - |01\rangle - |10\rangle + |11\rangle\big)\Big). \tag{2.38}$$

Simplifying this expression, we get:

$$|\psi_2\rangle = \frac{1}{2}(2|00\rangle \ -2|11\rangle) = |00\rangle - |11\rangle \tag{2.39}$$

Visualization of the Deutsch-Jozsa algorithm using a circuit diagram is shown in Figure 2.1, which depicts the quantum circuit involved in the algorithm.

2.5.2.11 Measurement

Finally, we measure the input qubits. If we obtain the state $|00\rangle$ (all zeros), the function is constant. Conversely, if we obtain the state $|11\rangle$ (all ones), the function is balanced.

2.5.2.12 The Algorithm Analysis

The Deutsch-Jozsa algorithm provides a significant speedup compared to classical approaches. Classically, determining the nature of the function would require an average of $2^{(n-1)} + 1$ evaluations for balanced functions. In contrast, the Deutsch-Jozsa algorithm solves the problem with a single query. This exponential improvement arises due to interference in the quantum superposition. If the function is constant, the interference of the amplitudes ensures destructive interference, resulting in a measurement outcome of $|00\rangle$. Conversely, if the function is balanced, the amplitudes interfere constructively, leading to a measurement outcome of $|11\rangle$ [23].

2.5.2.13 The Implementation Considerations

Implementing the Deutsch-Jozsa algorithm requires careful construction of the oracle transformation U_f, which encodes the function into the quantum circuit. The design of U_f depends on the specific function to be evaluated. In practice, this can be achieved through various quantum gate operations. The Deutsch-Jozsa algorithm demonstrates the power of quantum computation in solving a specific type of problem exponentially faster than classical methods. While the algorithm may not have direct practical applications, it serves as a foundational algorithm in understanding the potential of quantum computing. By harnessing the principles of superposition and interference, the Deutsch-Jozsa algorithm exemplifies the capabilities of quantum computing and paves the way for more advanced quantum algorithms with practical implications [23].

2.5.2.14 Bernstein-Vazirani Algorithm

The Bernstein-Vazirani algorithm is a quantum algorithm that efficiently solves the classical problem of the hidden string problem. Proposed by Ethan Bernstein and Umesh Vazirani in 1993, this algorithm showcases the power of quantum computation in solving certain types of problems with significant speedup compared to classical algorithms. In this section, we dive deep into the workings of the Bernstein-Vazirani algorithm, providing a comprehensive explanation along with the necessary mathematical details [24].

2.5.2.15 The Problem

The hidden string problem involves determining an unknown bit string by querying an oracle that provides access to the bits of the hidden string. The objective is to find the hidden string by making as few queries as possible. Mathematically, given a hidden string $s \in \{0, 1\}^n$ and an oracle function f_s: $\{0, 1\}^n \rightarrow \{0, 1\}$, the task is to find s using the fewest number of queries to the oracle f_s [24].

2.5.2.16 Oracle Function

In the context of the Bernstein-Vazirani algorithm, the oracle function f_s: $\{0, 1\}^n \rightarrow \{0, 1\}$ is a key component. It represents the unknown function or problem that the algorithm aims to solve. The oracle function takes as input an n-bit string and returns a single bit, providing access to the hidden string s.

The oracle function f_s is defined as follows:

$$f_s(x) = s \cdot x \pmod 2 \tag{2.40}$$

where x is an n-bit string, s is the hidden string we aim to determine, $\cdot$ represents the bitwise dot product, and (mod 2) ensures that the output is a single bit.

The oracle function evaluates the bitwise dot product of the hidden string s and the input string x. The dot product is computed by performing a bitwise AND operation between the corresponding bits of s and x, and then summing up the results. By querying the oracle with carefully chosen input strings, the algorithm can extract the entire hidden string s efficiently. Finally, the result is taken modulo 2 to obtain a single bit.

The power of the Bernstein-Vazirani algorithm lies in its ability to determine the hidden string s in a single query to the oracle, regardless of the size of n. This is achieved by utilizing the quantum properties of superposition and interference, allowing the algorithm to obtain information about all the bits of s simultaneously.

By querying the oracle function with a carefully chosen set of input strings, the Bernstein-Vazirani algorithm can unveil the hidden string s with remarkable efficiency, showcasing the power of quantum computation in solving certain types of problems with a significant speedup over classical approaches.

2.5.2.17 The Classical Approach

In the classical setting, determining the hidden string typically requires querying the oracle for each bit individually, resulting in n queries. The classical approach cannot solve this problem with fewer queries since accessing each bit separately is necessary [24]. Thus, the classical algorithm requires a linear number of queries, making it time-consuming for large problem instances [24].

2.5.2.18 The Quantum Approach

The Bernstein-Vazirani algorithm utilizes quantum computation to solve the hidden string problem with a single query, providing an exponential speedup compared to classical algorithms. The algorithm consists of the following steps:

Step 1: Initialization
Prepare a quantum register of n qubits in the state $|0^n\rangle$ and an additional qubit in the state $|1\rangle$.

Step 2: Superposition
Apply a Hadamard gate (H) to each qubit in the quantum register, creating a superposition of all possible input states:

$$|\psi_1\rangle = \frac{1}{\sqrt{2^n}} \sum_x |x\rangle, \tag{2.41}$$

where x represents a binary string of length n.

Step 3: Oracle Query
Apply the oracle function f_s to the quantum register by performing a controlled-NOT (CNOT) operation between each qubit in the register and the additional qubit, with the hidden string s as the control. This step encodes the hidden string information into the quantum register.

Step 4: Measurement
Apply a Hadamard gate (H) to each qubit in the quantum register, excluding the additional qubit. Then, perform measurements on each qubit in the register.

Step 5: Extraction
The measurements yield the hidden string s, which is determined by the observed values of the qubits in the quantum register.

2.5.2.19 Algorithm Execution

To better understand the workings of the Bernstein-Vazirani algorithm, let's consider an example of finding a hidden string s = "1101" using a quantum register of 4 qubits.

2.5.2.20 Initialization

We initialize a quantum register with 4 qubits in the state $|0000\rangle$ and an additional qubit in the state $|1\rangle$.

2.5.2.21 Superposition

Applying Hadamard gates to each qubit in the quantum register, we create a superposition of all possible input states:

$$|\psi_1\rangle = \frac{1}{2^2}\left(|0000\rangle + |0001\rangle + |0010\rangle + \cdots + |1111\rangle\right). \tag{2.42}$$

2.5.2.22 Oracle Query

We apply the oracle function f_s to the quantum register by performing controlled-NOT (CNOT) operations. In this case, the oracle encodes the hidden string s = "1101" into the quantum register.

2.5.2.23 Measurement

Applying Hadamard gates to each qubit in the quantum register (excluding the additional qubit), we obtain the final state:

$$|\psi_2\rangle = \frac{1}{2^2}\left(|0 \oplus s_0\rangle + |0 \oplus s_1\rangle + |0 \oplus s_2\rangle + |0 \oplus s_3\rangle\right) \tag{2.43}$$

where $\oplus$ represents bitwise addition modulo 2.

Measuring each qubit in the quantum register yields the hidden string s = "1101".

Visualization of the Bernstein-Vazirani algorithm using a circuit diagram illustrating the implementation of the Bernstein-Vazirani algorithm is shown in Figure 2.2, representing the quantum circuit for solving the hidden string problem.

2.5.2.24 Mathematical Representation

To represent the steps of the Bernstein-Vazirani algorithm mathematically, we can denote the initial state of the quantum register as $|\psi_0\rangle = |0^n 1\rangle$, where $|0^n\rangle$ represents the n-qubit state with all qubits initialized to $|0\rangle$, and $|1\rangle$ represents the additional qubit in state $|1\rangle$ [24]. The superposition state after applying Hadamard gates is represented as:

$$|\psi_1\rangle = \frac{1}{2^{n/2}} \sum_x |x\rangle \tag{2.44}$$

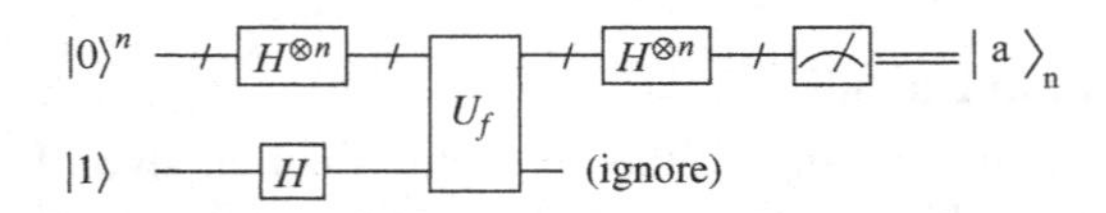

FIGURE 2.2 Circuit diagram of the Bernstein-Vazirani algorithm.

where Σ_x represents the summation over all binary strings x of length n. The oracle query can be represented as:

$$|\psi_2\rangle = \frac{1}{2^{n/2}} \sum_x (-1)^{(s \cdot x)} |x\rangle \tag{2.45}$$

where represents the bitwise dot product modulo 2 between s and x.

2.5.2.25 Example Application

To illustrate the concept of the Bernstein-Vazirani algorithm, let's consider a scenario involving three individuals: Alice, Bob, and Rudy.

Alice possesses a hidden string s = "1010". Bob and Rudy want to determine the hidden string by querying Alice, but they can only communicate via a quantum channel.

Using the Bernstein-Vazirani algorithm, Bob can obtain the hidden string by querying Alice's quantum oracle only once. The algorithm encodes the hidden string information into the quantum register, and Bob can extract the hidden string by performing measurements on the qubits.

By following the steps of the algorithm, Bob successfully determines the hidden string s = "1010" without the need for multiple queries to Alice's oracle.

2.5.2.26 Summary

The Bernstein-Vazirani algorithm demonstrates the power of quantum computation in solving the hidden string problem with a single query, providing an exponential speedup compared to classical algorithms. By leveraging quantum superposition and interference, the algorithm efficiently extracts the hidden string information from a quantum register. The mathematical framework and steps of the algorithm illustrate the fundamental principles of quantum computation and its potential for solving complex problems.

2.5.3 Grover's Algorithm

Grover's algorithm, developed by Lov Grover in 1996, is a groundbreaking quantum algorithm that offers a quadratic speedup for searching unsorted databases. By harnessing the principles of quantum superposition and interference, Grover's algorithm provides an efficient solution to the search problem. Unlike classical search algorithms that require examining each item individually, Grover's algorithm can quickly locate a target item with a complexity of $O(\sqrt{N})$, where N is the size of the database. This algorithm has far-reaching implications in various fields, including database searching, optimization, and cryptography, showcasing the remarkable capabilities of quantum computing [25].

2.5.4 The Problem

In the realm of information retrieval, efficient search algorithms play a vital role in locating specific items within large, unsorted databases. Classical search methods, such as linear or binary search, require sequential examination of each item, resulting in a time complexity of $O(N)$ for a database of size N. However, as the size of databases grows exponentially, these traditional approaches become increasingly inefficient and time-consuming.

To overcome this challenge, there is a pressing need for a breakthrough algorithm that can offer a significant speedup in the search process. Grover's algorithm, a remarkable quantum algorithm, presents a compelling solution to this problem. By harnessing the principles of quantum mechanics, Grover's algorithm demonstrates the potential to revolutionize information retrieval by providing a quadratic speedup over classical algorithms. The problem at hand is to design and implement Grover's algorithm to efficiently search an unsorted database and identify the target item with optimal time complexity. This entails developing a quantum circuit that can leverage quantum superposition and interference to amplify the probability of finding the target item. By exploring the mathematical foundations, quantum circuit design, and potential applications of Grover's algorithm, we aim to unlock the transformative power of quantum computing in information retrieval [25].

The successful implementation of Grover's algorithm will have profound implications across diverse domains, including database management, optimization problems, and cryptographic key searching. By addressing the limitations of classical search algorithms and harnessing the power of quantum mechanics, Grover's algorithm offers a promising avenue for enhancing search efficiency and accelerating information retrieval processes.

2.5.5 The Classical Approach

Traditionally, the search for a specific item within an unsorted database has relied on classical algorithms that sequentially examine each element. The most basic approach is the linear search, which starts from the first item and proceeds through the database until a match is found or the end is reached. This method has a time complexity of $O(N)$, where N represents the size of the database. To improve efficiency, binary search can be employed when the database is sorted. Binary search divides the database into halves and compares the target item with the middle element, eliminating half of the remaining search space with each iteration. This approach achieves a time complexity of $O(\log N)$, which is significantly faster than linear search for large databases.

However, both linear and binary search algorithms display limitations when faced with unsorted databases. Linear search requires examining every item, resulting in a worst-case scenario of having to search through the entire database. Binary search, on the other hand, relies on the database being sorted, which may not always be the case. As the size of databases continues to grow exponentially, these classical search methods become increasingly inefficient and time-consuming. Therefore, there is a clear need for an alternative approach that can provide a substantial speedup and overcome the limitations of classical algorithms in unsorted databases [25].

2.5.6 The Quantum Approach

Grover's algorithm offers a revolutionary quantum approach to the problem of searching unsorted databases. By harnessing the power of quantum superposition and interference, this algorithm can efficiently locate the target item with a quadratic speedup compared to classical algorithms. The core idea behind Grover's algorithm is to exploit quantum parallelism to explore multiple database entries simultaneously, leading to a more efficient search process. At the heart of Grover's algorithm lies the principle of quantum amplitude amplification. It capitalizes on the ability of quantum systems to be in a superposition of multiple states simultaneously. By applying a sequence of quantum operations, including quantum oracles and quantum diffusion operators, Grover's algorithm amplifies the amplitude of the target item and suppresses the amplitudes of non-target items.

Implementing Grover's algorithm encodes the items in the database into quantum states. This encoding is typically achieved through a process known as amplitude encoding, which maps each item to a quantum state with a specific amplitude. The target item is assigned a higher amplitude, while the non-target items have lower amplitudes. This encoding scheme is crucial for the success of Grover's algorithm. The quantum oracle is constructed to mark the target item. The oracle performs a phase inversion operation on the amplitude of the target item, effectively flipping its sign. This transformation enhances the amplitude of the target item, making it distinguishable from the other items in the database. After marking the target item, a quantum diffusion operator is employed to spread the amplitude across all database entries. This operator serves to increase the amplitudes of the non-target items and decrease the amplitude of the target item. By iteratively applying the quantum oracle and quantum diffusion operator, Grover's algorithm converges toward the target item [25].

2.5.7 The Algorithm Execution

The execution of Grover's algorithm involves a series of steps that exploit quantum parallelism and interference to efficiently search for a target item within an unsorted database. Let's delve into each step in detail.

2.5.8 Initialization

Initialize a quantum register of n qubits in the state $|\psi\rangle = |0\rangle^{\otimes n}$, where $|0\rangle$ represents the base state, representing the search space of size $N = 2^n$.

2.5.9 Superposition

Apply a Hadamard gate (H) to each qubit in the quantum register to create a superposition of all possible states:

$$|\psi_1\rangle = \frac{1}{\sqrt{2^n}} \sum_x |x\rangle. \tag{2.46}$$

2.5.10 Oracle Query

The oracle function marks the target item in the database. It acts as a black box, inverting the phase of the target item while leaving the other items unchanged.

Construct an oracle function that marks the desired solution(s). The oracle transforms the amplitude of the target item(s) to −1, while leaving the other items unchanged. The oracle function can be represented as $U_f|x\rangle = (-1)^{f(x)}|x\rangle$, where $f(x) = 1$ if x is the target item and $f(x) = 0$ otherwise.

2.5.11 Amplitude Amplification

Apply the amplitude amplification operator (A) to enhance the amplitude of the target item in the superposition. The amplification process involves repeating the inversion about the mean (I) and the oracle query (U_f) operations $\sqrt{N}$ times, where N is the size of the database. Each iteration consists of two steps: reflection about the mean and the application of the oracle function.

2.5.12 Reflection about the Mean

Perform an inversion operation about the mean state, denoted as $D = 2|s\rangle\langle s| - I$, where I is the identity matrix. This reflection flips the amplitudes around the mean, effectively concentrating the probability amplitudes of the target item(s). Mathematically, the reflection operation can be represented as $D = 2|s\rangle\langle s| - I$.

2.5.13 Oracle Function

Apply the oracle function U_f to mark the target item(s) with a phase inversion. This operation effectively enhances the probability amplitude of the target item(s) in subsequent iterations. Mathematically, the application of the oracle function can be represented as $U_f = I - 2|w\rangle\langle w|$, where $|w\rangle$ is the solution state.

2.5.14 Measurement

Measure the qubits in the quantum register. The probability of obtaining the index of the target item is significantly increased due to the amplification process. Repeating the algorithm $O(\sqrt{N})$ times increases the chances of obtaining the correct result with high probability. Visualization of the Grover's algorithm using a circuit diagram is shown in Figure 2.3, representing the quantum circuit for searching unsorted databases.

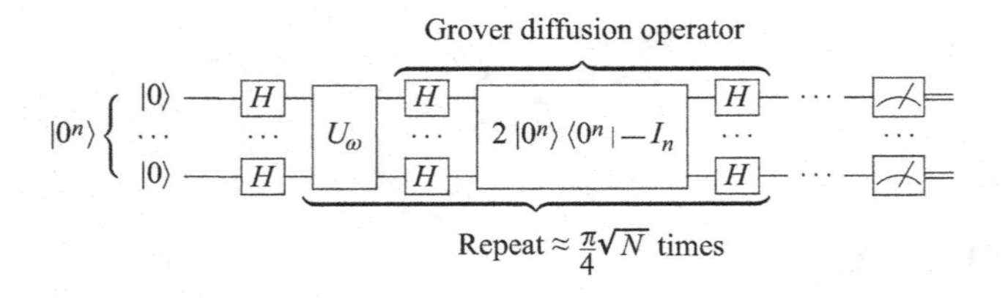

FIGURE 2.3 Circuit diagram of Grover's algorithm.

2.5.15 The Example Application

To illustrate the workings of Grover's algorithm, let's consider a scenario involving three individuals: Alice, Bob, and Rudy. Alice has a secret number in mind, and Bob's task is to guess the number by asking yes-or-no questions. Rudy, on the other hand, will assist Bob in finding the correct number using Grover's algorithm.

In the classical approach, Bob would need to ask a series of questions one by one until he stumbles upon the correct answer. For example, he might start by asking if the number is less than five, to which Alice would respond "no." Bob would continue asking more questions, narrowing down the possibilities until he finally arrives at the correct number. Now, let's introduce Grover's algorithm to the mix. Rudy sets up a quantum system that represents all possible numbers, with each number encoded as a quantum state. Initially, all these states are in a superposition, meaning they exist simultaneously. The goal is to amplify the amplitude of the state representing Alice's secret number while reducing the amplitudes of the other states.

Rudy starts by applying a series of quantum operations to the quantum system, including a reflection about the mean and an oracle function. The reflection about the mean flips the amplitudes around the average, concentrating the probability amplitudes of the target state. The oracle function marks Alice's secret number with a phase inversion, enhancing its probability amplitude. Through multiple iterations of applying the reflection and oracle function, Rudy increases the probability of measuring Alice's secret number. Eventually, after a sufficient number of iterations, Rudy performs a measurement on the quantum system. When Bob receives the measurement result, he has a high probability of obtaining Alice's secret number, thanks to Grover's algorithm. This quantum approach provides a significant speedup compared to the classical method, as fewer questions need to be asked to arrive at the correct answer.

2.5.16 Summary

Grover's algorithm is a powerful quantum search algorithm that provides a quadratic speedup over classical search methods. By leveraging the principles of quantum mechanics, it offers an efficient solution to finding a target item within an unsorted database. The algorithm achieves this by employing quantum superposition and interference, enabling a significant reduction in time complexity.

The time complexity of Grover's algorithm is approximately $O(\sqrt{N})$, where N represents the size of the search space. This is a substantial improvement compared to the classical counterpart, which typically requires $O(N)$ operations. As a result, Grover's algorithm offers a remarkable speedup, especially for large databases. The space complexity of Grover's algorithm remains unchanged compared to classical search methods, requiring $O(n)$ qubits to represent the search space of size $N = 2^n$. This space requirement is feasible for small to moderate problem sizes but can become challenging for larger search spaces due to the physical limitations of quantum hardware.

One of the limitations of Grover's algorithm is that it does not provide a direct solution to optimization problems or problems with complex constraints. It is primarily designed for the search problem in unsorted databases. Additionally, the algorithm relies on an accurate oracle function that marks the target item(s) efficiently, which may pose challenges in certain problem domains.

Despite its limitations, Grover's algorithm has real-life applications in various domains. For example, it can be used for data-searching and pattern-matching tasks in databases, cryptographic applications such as key searching and brute-force attacks, and optimization problems where finding the global minimum or maximum is required.

2.5.17 Shor's Algorithm

Shor's algorithm, proposed by Peter Shor in 1994, is a groundbreaking quantum algorithm that efficiently factors large composite numbers, challenging the limitations of classical computers. This algorithm has significant implications for cryptography, as many encryption schemes rely on the difficulty of factoring large numbers. Shor's algorithm demonstrates the potential of quantum computing to outperform classical computers in solving complex mathematical problems [26].

2.5.18 The Problem

The problem that Shor's algorithm addresses is the factorization of large composite numbers, which involves finding the prime factors that multiply together to yield the given number. Factorization is considered a difficult problem for classical computers, as the best-known classical algorithms have exponential time complexity. This problem underpins the security of many cryptographic systems, as the difficulty of factorization is exploited in key exchange protocols. Shor's algorithm aims to efficiently factorize large composite numbers using quantum computations, thereby threatening the security of classical cryptographic systems [26].

2.5.19 The Classical Approach

In the classical approach to factorization, methods such as trial division, Fermat's factorization, and Pollard's rho algorithm are commonly employed. These classical algorithms are suitable for small numbers but become increasingly inefficient as the size of the number to be factored grows. They have exponential time complexity, making them impractical for factoring large numbers.

Among these classical algorithms, the general number field sieve (GNFS) stands out as the most efficient method for factoring large composite numbers. The GNFS has a sub-exponential time complexity of $O\left(\exp\left(\left(\frac{64}{9}\right)^{1/3}(\ln N)^{1/3}(\ln N)^{2/3}\right)\right)$, where N represents the number to be factored. This makes the GNFS significantly faster than other classical algorithms for large numbers.

The GNFS combines several mathematical techniques, including sieving, linear algebra, and trial division, to identify the prime factors of a composite number.

It begins with a sieving step, where a large range of integers is examined to identify values that potentially share common factors with the target number. The sieving process relies on finding smooth numbers, which are integers with small prime factors. These smooth numbers are then used in the subsequent linear algebra step to solve a system of linear equations and find linear dependencies among the rows. After the linear algebra step, the GNFS applies a square rooting technique to find a congruence relation that reveals information about the prime factors. Finally, a trial division is performed on the remaining factors to verify their primality and complete the factorization process. While the GNFS is a significant improvement over earlier classical algorithms, it still remains computationally expensive for large numbers. The sub-exponential time complexity implies that the GNFS grows faster than a polynomial but slower than an exponential with respect to the input size. As a result, the GNFS becomes increasingly impractical as the size of the number to be factored increases.

In contrast, Shor's algorithm, a quantum algorithm for factoring large numbers, offers a remarkable speedup compared to classical methods. Shor's algorithm can factorize large numbers efficiently on a quantum computer, leveraging the principles of quantum mechanics, such as superposition and entanglement. This exponential speedup in factoring large numbers is one of the reasons why quantum computers pose a significant threat to classical cryptography [26].

2.5.20 The Quantum Approach

The quantum approach in Shor's algorithm consists of several key steps. It begins with the initialization of quantum registers to store the input number and an auxiliary register. The input register is prepared in a superposition of all possible inputs, allowing for simultaneous exploration of solutions. Modular exponentiation is then performed using a unitary operator, followed by the application of the quantum Fourier transform (QFT) to the input register. Measurement is performed, and the measurement outcome is subjected to the continued fraction algorithm, revealing the factors of the input number. The significance of the quantum approach in Shor's algorithm lies in its potential to break classical cryptographic systems that rely on the difficulty of factoring large numbers. If successfully implemented, it could compromise widely used encryption methods and have far-reaching implications for data security. However, there are significant challenges in practical implementation, including the requirement for error-corrected qubits and fault-tolerant quantum operations, as well as the scalability of the algorithm to factor significantly larger numbers.

This approach in Shor's algorithm consists of several key components that work together to factor large numbers. The algorithm begins by initializing two quantum registers: the input register and an auxiliary register. The input register represents the number to be factored and is prepared in a superposition of all possible inputs, allowing for simultaneous exploration of multiple solutions [26].

The next step involves modular exponentiation, where a unitary operator performs exponentiation on the input register in a modular arithmetic framework. This step extracts periodicity information from the quantum state, crucial for finding the factors efficiently. Following modular exponentiation, the QFT is applied to the input register. The QFT is a quantum analog of the classical Fourier transform and enables the transformation of the quantum state into a representation that reveals the periodic structure of the input number. After the QFT, a measurement is performed on the quantum state, collapsing it into a classical state and providing a specific outcome. This outcome is essential for determining the factors of the input number. The final step involves subjecting the measurement outcome to the continued fraction algorithm. This algorithm analyzes the measurement result using mathematical techniques, such as continued fraction expansions, to efficiently extract the factors of the input number.

2.5.21 The Algorithm Execution

2.5.21.1 Initialization

To begin the algorithm, we prepare two quantum registers: the input register and the auxiliary register. The input register contains the number to be factored, denoted as N. The auxiliary register is initialized to the state $|0\rangle$. We can represent the initial state of the input register as $|\psi\rangle = |1\rangle$.

2.5.21.2 Superposition

In this step, we create a superposition of all possible inputs in the input register. We achieve this by applying a series of Hadamard gates to the input register. The Hadamard gate, denoted as H, transforms the basis states as follows: $H|0\rangle = \frac{1}{\sqrt{2}}(|0\rangle + |1\rangle)$ and $H|1\rangle = \frac{1}{\sqrt{2}}(|0\rangle - |1\rangle)$. Applying the Hadamard gate to the input register, we obtain the superposition state:

$$|\psi\rangle = \frac{1}{\sqrt{M}} \sum_{x=0}^{M-1} |x\rangle,$$

where x ranges from 0 to $M - 1$ and $M = 2^n$, with n being the number of qubits in the input register.

2.5.21.3 Modular Exponentiation

Modular exponentiation is a crucial step in Shor's algorithm that involves raising a base, denoted as "a," to the power of an exponent, denoted as "x," and then computing the result modulo N. The goal is to efficiently calculate the value of (a^x) mod N.

The modular exponentiation step is implemented using a unitary operator called U_f, which performs the mapping $|x\rangle|y\rangle \rightarrow |x\rangle|a^y \bmod N\rangle$. Here, $|x\rangle$ represents the state of the input register, $|y\rangle$ represents the state of the auxiliary register, and "a_y mod N" represents the modular exponentiation of "a" raised to the power of "y" modulo N.

To understand the implementation of modular exponentiation, we can consider a simple example. Let's assume we want to calculate (3^7) mod 10. Here, the base "a" is 3, the exponent "x" is 7, and N is 10.

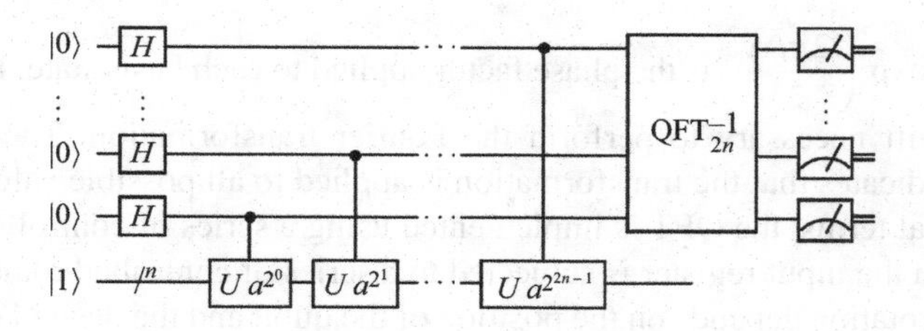

FIGURE 2.4 Circuit diagram of Shor's algorithm.

2.5.21.4 Applying the Unitary Operator U_f

To perform modular exponentiation, we apply the unitary operator U_f to the state $|x\rangle\,|y\rangle$. In this case, Uf maps $|x\rangle\,|y\rangle$ to $|x\rangle\,|a^y \bmod N\rangle$.

For our example, let's consider the value of "a" as 3. Applying U_f, we get:

$$U_f : |x\rangle|y\rangle \rightarrow |x\rangle|(3^y) \bmod 10\rangle.$$

The auxiliary register gets updated based on the modular exponentiation of "a" raised to the power of "y" modulo N.

Visualization of the Shor's algorithm using a circuit diagram is shown in Figure 2.4, representing the quantum circuit for factoring large composite numbers.

2.5.21.5 Quantum Fourier Transform (QFT)

The quantum Fourier transform (QFT) is a fundamental component of Shor's algorithm. It is a quantum analogue of the classical discrete Fourier transform (DFT) and plays a crucial role in identifying the periodicity of the modular exponentiation function.

The QFT operates on an input register of n qubits, where the qubits are in a superposition state $|x\rangle$. The goal of the QFT is to transform the input register from the computational basis to the Fourier basis, allowing for efficient analysis of the periodicity introduced by the modular exponentiation step [27].

The QFT can be defined mathematically as follows:

$$\text{QFT} : |x\rangle \rightarrow \frac{1}{\sqrt{2^n}} \sum_{y=0}^{2^n-1} \exp\left(\frac{2\pi ixy}{2^n}\right) |y\rangle. \tag{2.47}$$

Let's break down the mathematical expression and understand its significance:

$|x\rangle$ represents the input state of the qubits in the computational basis. It is a superposition of all possible values of x, where x ranges from 0 to $2^n - 1$. $|y\rangle$ represents the transformed state of the qubits in the Fourier basis. It is a superposition of all possible values of y, where y ranges from 0 to $2^n - 1$. The term $\frac{1}{\sqrt{2^n}}$ normalizes the state to ensure that the total probability of all possible outcomes sums to 1.

The term $\exp\left(\frac{2\pi ixy}{2^n}\right)$ is the phase factor applied to each basis state. It introduces the phase shift necessary to perform the Fourier transformation. The summation symbol Σ_y indicates that the transformation is applied to all possible values of y.

In practical terms, the QFT is implemented using a series of controlled rotations. Each qubit in the input register is subjected to a series of controlled rotations, where the angle of rotation depends on the position of the qubit and the desired output state. The QFT provides a powerful tool for identifying the periodicity introduced by the modular exponentiation step. By transforming the input register to the Fourier basis, the QFT allows for the extraction of information about the period of the function, which is crucial in determining the factors of the number being factored.

2.5.21.6 Measurement and Continued Fraction Algorithm

After applying the QFT to the output of the modular exponentiation step, we obtain a superposition of states. To extract useful information from this superposition, we need to measure the quantum state, obtaining a classical outcome. The measurement outcome is a specific value, denoted as y, which represents the result of the measurement in the Fourier basis. This measurement is a crucial step as it provides us with the information needed to determine the factors of the number N.

Next, we employ the continued fraction algorithm to analyze the measurement outcome. The goal is to obtain rational approximations that reveal the period of the measured value and, subsequently, the factors of N. The continued fraction algorithm iteratively constructs a sequence of fractions, each representing an approximation of the original measurement outcome y divided by the total number of possible outcomes, M. The sequence starts with the fraction $\frac{y}{M}$ [28].

2.5.21.7 The Example Application

Alice and Bob are good friends who are collaborating to factorize a large number, N. Here's how they work together:

Step 1: Preparation
Alice selects the number she wants to factorize, N. For this example, let's assume $N = 35$.

Step 2: Quantum Circuit Preparation
Alice and Bob work together to design the quantum circuit required for Shor's algorithm. They determine the modular exponentiation step, quantum Fourier transform, and measurement operations needed for the chosen value of N.

Step 3: Modular Exponentiation
Alice instructs Bob to perform the modular exponentiation step. They choose a random base, a, that is coprime to N. Let's say they select $a = 2$. Bob implements the modular exponentiation circuit by applying the unitary operator Uf, which maps $|x\rangle\,|y\rangle$ to $|x\rangle\,|(a^y \bmod N)\rangle$. This involves performing modular multiplications, raising a to various powers modulo N.

For instance, if Bob measures the auxiliary register and obtains the outcome $|y\rangle = 3$, he computes (a^y mod N) = (2^3 mod 35) = 8 mod 35. Bob communicates this result, 8, back to Alice.

Step 4: Quantum Fourier Transform

Alice instructs Bob to apply the quantum Fourier transform (QFT) on the input register $|x\rangle$. The QFT transforms the state of the input register from the computational basis to the Fourier basis, allowing for efficient analysis of the periodicity introduced during the modular exponentiation step.

For example, if the measurement outcome after the QFT is $|x\rangle = 6$, it represents a superposition of possible values, each associated with a different periodicity related to the factors of N.

Step 5: Measurement and Continued Fraction Algorithm

Alice instructs Bob to measure the output of the QFT. They obtain a measurement outcome, let's say $|x\rangle = 6$. Alice and Bob then apply the continued fraction algorithm to approximate the value of y/M, where y represents the measurement outcome and M is the total number of possible outcomes.

Using the continued fraction algorithm, they find that the rational approximation for $6/M$ is 1/6. This approximation reveals the presence of a hidden periodicity in the measurement outcome, which is crucial for determining the factors of N.

Step 6: Factors Extraction

Alice and Bob work together to extract the factors of N using the information obtained from the continued fraction algorithm. In this case, the continued fraction approximation of 1/6 indicates a periodicity that corresponds to the factors 5 and 7. Bob communicates the factors back to Alice, who verifies them. They have successfully factorized the large number $N = 35$ into its prime factors: 5 and 7.

2.5.21.8 Summary

Shor's algorithm is a groundbreaking quantum algorithm that revolutionizes the field of factorization and has far-reaching implications for cryptography and number theory. By leveraging the power of quantum computing, Shor's algorithm offers an exponential speedup compared to classical algorithms, enabling efficient factorization of large numbers.

Shor's algorithm has significant implications for cryptography. The security of widely used cryptographic protocols, such as RSA, relies on the difficulty of factorizing large numbers. Shor's algorithm poses a significant threat to these cryptographic systems, as it can factorize large numbers efficiently. This has led to an increased interest in post-quantum cryptography, which focuses on developing new encryption schemes that are resistant to attacks from quantum computers. Shor's algorithm has practical applications in other areas such as computational number theory and optimization. It can be used to solve problems that are intrinsically related to factorization, such as finding the period of a periodic function and solving certain types of Diophantine equations. While Shor's algorithm offers groundbreaking capabilities, it is important to note its limitations. The main limitation is the requirement for a fully functional, error-corrected quantum computer to execute the algorithm efficiently. Building and maintaining such a quantum computer at a large scale is currently a significant technological challenge.

The time complexity of Shor's algorithm can be expressed as $O((\log N)^3)$, where N is the number to be factorized. This time complexity is significantly better than the best-known classical algorithms, which have exponential time complexity. Shor's algorithm offers an exponential speedup, enabling the efficient factorization of large numbers that would be infeasible with classical methods. The time complexity mentioned assumes a fully functional, error-corrected quantum computer. However, practical implementations of Shor's algorithm on currently available quantum hardware face various challenges, including decoherence, noise, and limited qubit resources. As a result, the current implementations of Shor's algorithm are still limited to relatively small numbers.

2.6 CONCLUSION

This chapter delves into the transformative realm of quantum computing, showcasing its potential to reshape computational capabilities fundamentally. Beginning with an introduction to quantum machines and their distinctive features, the narrative progresses to explore fundamental quantum computing concepts such as qubits, superposition, and entanglement, elucidating the foundational role of quantum gates and operations in constructing quantum circuits. A comprehensive understanding of quantum mechanics, encompassing quantum states, wavefunctions, and measurement processes, is essential for harnessing the power of quantum computing. The chapter elucidates the representation and notation of quantum bits and states, delving into Hilbert's space and the Bloch sphere for visualization. Quantum gates and circuits, including single-qubit and two-qubit gates, are examined, with a focus on their significance in quantum circuit design. The exploration culminates in an in-depth analysis of prominent quantum algorithms, including the Deutsch-Jozsa, Bernstein-Vazirani, Grover's, and Shor's algorithms, offering step-by-step implementation guides. By laying this foundation, the chapter sets the stage for future research and applications of quantum algorithms, foreseeing their potential to revolutionize fields like cryptography, optimization, and data searching as the unlocking of quantum computing's power continues, opening up unprecedented possibilities for efficient solutions to complex problems. In the realm of future research directions, quantum computing stands poised to lead a transformative era in computational science. As the field continues to evolve, one promising avenue is the development of fault-tolerant quantum computers. Mitigating the impact of quantum errors and achieving fault tolerance is a critical challenge that researchers are actively addressing, as it will pave the way for more reliable and scalable quantum processors.

REFERENCES

[1] Andrew Steane. Quantum computing. *Reports on Progress in Physics*, 61(2):117, 1998.

[2] Marius Nagy and Selim G Akl. Quantum computing: Beyond the limits of conventional computation. *The International Journal of Parallel, Emergent and Distributed Systems*, 22(2):123–135, 2007.

[3] Albert Messiah. *Quantum Mechanics*. Courier Corporation, 2014.
[4] Freeman J Dyson and David Derbes. *Advanced Quantum Mechanics*. World Scientific, 2011.
[5] Yuli V Nazarov and Jeroen Danon. *Advanced Quantum Mechanics: A Practical Guide*. Cambridge University Press, 2013.
[6] Guanru Feng, Dawei Lu, Jun Li, Tao Xin, and Bei Zeng. Quantum computing: principles and applications. *arXiv preprint arXiv:2310.09386*, 2023.
[7] Pradosh K Roy. *Fundamentals of Quantum Computing Part I*. 2020. DOI:10.13140/RG.2.2.34231.14242
[8] Matteo GA Paris. The modern tools of quantum mechanics: A tutorial on quantum states, measurements, and operations. *The European Physical Journal Special Topics*, 203(1):61–86, 2012.
[9] Bernard Zygelman. *A First Introduction to Quantum Computing and Information*. Springer, 2018.
[10] David P DiVincenzo. Quantum gates and circuits. *Proceedings of the Royal Society of London. Series A: Mathematical, Physical and Engineering Sciences*, 454(1969):261–276, 1998.
[11] Colin P Williams and Colin P Williams. Quantum gates. *Explorations in Quantum Computing*, pp. 51–122, 2011.
[12] Giulio Chiribella, G Mauro D'Ariano, and Paolo Perinotti. Quantum circuit architecture. *Physical Review Letters*, 101(6):060401, 2008.
[13] Katherine L Brown, William J Munro, and Vivien M Kendon. Using quantum computers for quantum simulation. *Entropy*, 12(11):2268–2307, 2010.
[14] Andrew J Daley, Immanuel Bloch, Christian Kokail, Stuart Flannigan, Natalie Pearson, Matthias Troyer, and Peter Zoller. Practical quantum advantage in quantum simulation. *Nature*, 607(7920):667–676, 2022.
[15] Andreas Bayerstadler, Guillaume Becquin, Julia Binder, Thierry Botter, Hans Ehm, Thomas Ehmer, Marvin Erdmann, Norbert Gaus, Philipp Harbach, Maximilian Hess, et al. Industry quantum computing applications. *EPJ Quantum Technology*, 8(1):25, 2021.
[16] Naya-Plasencia María. Post-quantum symmetric cryptography. *Symmetric Cryptography, Volume 2: Cryptanalysis and Future Directions*, p. 203, Wiley, 2024.
[17] Claudio Conti. *Quantum Machine Learning: Thinking and Exploration in Neural Network Models for Quantum Science and Quantum Computing*. Springer Nature, 2024.
[18] Bela Bauer, Sergey Bravyi, Mario Motta, and Garnet Kin-Lic Chan. Quantum algorithms for quantum chemistry and quantum materials science. *Chemical Reviews*, 120(22):12685–12717, 2020.
[19] Anshul Saxena, Javier Mancilla, Iraitz Montalban, and Christophe Pere. *Financial Modeling Using Quantum Computing: Design and Manage Quantum Machine Learning Solutions for Financial Analysis and Decision Making*. Packt Publishing Ltd, 2023.
[20] Yu-Bo Sheng, Lan Zhou, and Gui-Lu Long. One-step quantum secure direct communication. *Science Bulletin*, 67(4):367–374, 2022.
[21] Oktay K Pashaev. Quantum coin flipping, qubit measurement, and generalized Fibonacci numbers. *Theoretical and Mathematical Physics*, 208(2):1075–1092, 2021.
[22] Francesco Tacchino, Alessandro Chiesa, Stefano Carretta, and Dario Gerace. Quantum computers as universal quantum simulators: State-of-the-art and perspectives. *Advanced Quantum Technologies*, 3(3):1900052, 2020.
[23] Antonio N Oliveira, Estêvão VB de Oliveira, Alan C Santos, and Celso J Villas-Bôas. Quantum algorithms in IBMQ experience: Deutsch-jozsa algorithm. *arXiv preprint arXiv:2109.07910*, 2021.
[24] Xu Zhou, Daowen Qiu, and Le Lou. Distributed exact quantum algorithms for Bernstein-Vazirani and search problems. *arXiv preprint arXiv:2303.10670*, 2023.

[25] Bikram Khanal, Pablo Rivas, Javier Orduz, and Alibek Zhakubayev. Quantum machine learning: A case study of Grover's algorithm. In *2021 International Conference on Computational Science and Computational Intelligence (CSCI)*, pp. 79–84. IEEE, 2021.

[26] Hiu Yung Wong. Shor's algorithm. In *Introduction to Quantum Computing: From a Layperson to a Programmer in 30 Steps*, pp. 289–298. Springer, 2023.

[27] YS Nam and Reinhold Blümel. Scaling laws for Shor's algorithm with a banded quantum Fourier transform. *Physical Review A*, 87(3):032333, 2013.

[28] Johanna Barzen and Frank Leymann. Continued fractions and probability estimations in Shor's algorithm: a detailed and self-contained treatise. *Applied Mathematics*, 2(3):393–432, 2022.

Part II

Introduction to Quantum Machine Learning

3 Machine Learning with Supervised Quantum Models

Nisha Soms
KPR Institute of Engineering and Technology, India

David Samuel Azariya
Sona College of Technology, India

Savitha Selvi Jagathiswaramoorthi
Sri Ramakrishna Institute of Technology, India

Mohanraj Vijayakumar
Sona College of Technology, India

3.1 INTRODUCTION

Machine Learning (ML) with supervised quantum models is a cutting-edge field that combines the power of ML algorithms with the potential of quantum computing. This approach aims to leverage the unique properties of quantum systems to enhance the performance of supervised learning tasks. In this paradigm, quantum models are utilized as the underlying framework for data processing and analysis. By harnessing the principles of superposition and entanglement, these models can handle complex computations more efficiently than classical counterparts. This opens up new possibilities for solving intricate problems in various domains, such as optimization, pattern recognition, and data classification.

Through the use of supervised quantum models, quantum systems may be trained to learn from labeled datasets and generate precise predictions about yet-to-be-observed data. The learning potential and prospective accuracy of the models can be improved through the combination of classical ML approaches and quantum algorithms. It's crucial to keep in mind that supervised quantum machine learning is still in its infancy and still struggles with issues like noise and error correction in quantum systems. Nevertheless, recently discovered, at the intersection of quantum computing and machine learning, is the fascinating field of quantum machine learning [1] that will advance present machine learning approaches and improve management of

DOI: 10.1201/9781003429654-5

difficult issues. The advantages, challenges, and key concepts of several quantum machine learning methods are thoroughly reviewed in this chapter.

3.2 OVERVIEW OF QUANTUM MACHINE LEARNING

In a wide range of industries, ML algorithms have shown remarkable performance. The high computing complexity of many difficult issues, however, still necessitates the employment of traditional methods. Through the use of parallel quantum processing and quantum computing, quantum machine learning [2] offers the ability to get beyond these restrictions.

3.2.1 Key Goals

The principal goals of quantum ML algorithms include:

- Improving the effectiveness of classical ML algorithms [3] by utilizing quantum parallelism [4] and quantum superposition [5].
- Utilizing quantum algorithms and principles to solve inherently quantum problems, such as quantum state classification or data analysis.
- Developing hybrid algorithms that leverage the strengths of both classical machine learning and quantum computing, while exploring their synergy.

3.2.2 Benefits

- ***Speedup in computation***: Quantum computers offer the potential for exponential speedup in certain computations compared to classical counterparts. Quantum machine learning algorithms aim to harness this speedup to perform computations more efficiently, especially for problems with large datasets or complex feature spaces.
- ***Enhanced data processing***: Quantum machine learning algorithms can exploit quantum states and quantum operations to process and analyze data more effectively, enabling more accurate predictions and insights.
- ***Quantum feature space***: The idea of a quantum feature space [6] is introduced by quantum machine learning, allowing for richer data representations and offering possible benefits in solving challenging issues.

3.2.3 Challenges

- ***Quantum hardware limitations***: The advancement of quantum machine learning techniques is always linked to the accessibility and scalability of quantum technology. Scaling and running quantum algorithms is difficult due to the limitations on the number of qubits and the quality of quantum operations at the moment.
- ***Noise and error correction***: The dependability and accuracy of quantum machine learning techniques may be impacted by the noise and errors that these systems are prone to. To address these issues, it is essential to create methods for error correction and fault-tolerant quantum computing.

- ***Hybrid approaches***: Integrating quantum algorithms with classical machine learning techniques poses challenges in terms of algorithm design, data preprocessing, and model optimization. Research is still being done on effective hybrid methods that combine the best features of conventional and quantum systems.

3.3 OVERVIEW OF QUANTUM MACHINE LEARNING ALGORITHMS

Quantum machine learning algorithms are a class of algorithms that apply the principles of quantum computing to enhance the performance of traditional machine learning methodologies. These algorithms aim to exploit the unique properties of quantum systems, such as superposition and entanglement, to accelerate computations and, perhaps, solve difficult complex problems on classical computers. The following are some well-known quantum machine learning algorithms (see Figure 3.1).

The quantum ML algorithms, as mentioned above, can be categorized into distinct groups based on their underlying principles and techniques. The categorization is as follows:

1. **Supervised learning:** The quantum support vector machine (QSVM) algorithm falls under this category. It focuses on performing classification tasks by finding an optimal hyperplane to separate data points.
2. **Unsupervised learning:** This category includes the techniques quantum principal component analysis (QPCA) and quantum K-means clustering.

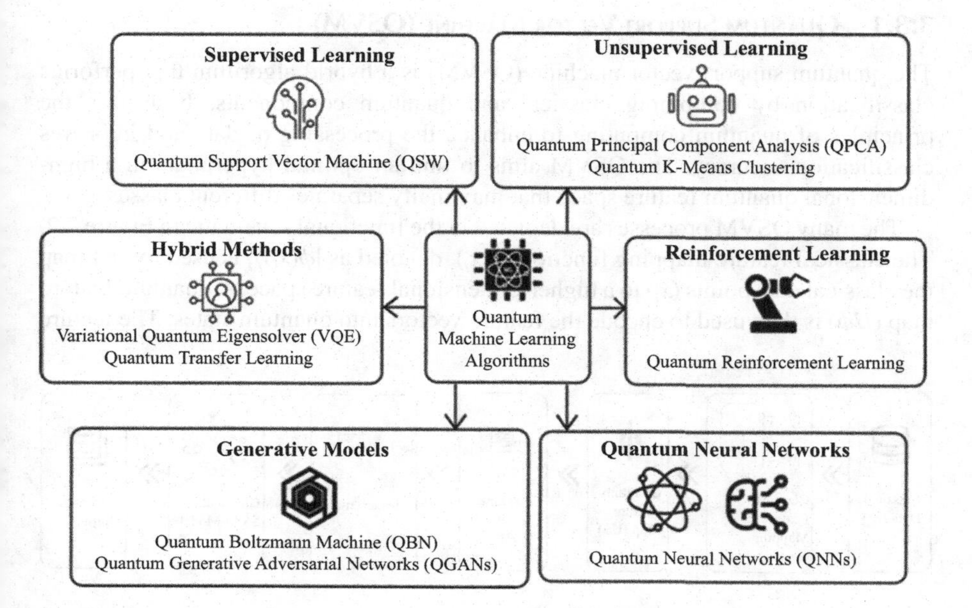

FIGURE 3.1 Quantum machine learning algorithms.

Quantum K-means clustering uses quantum circuits to combine like data points whereas QPCA uses quantum state mapping to extract the most important features from high-dimensional data.

3. **Generative models:** The quantum Boltzmann machine (QBM) and quantum generative adversarial networks (QGANs) fall under this category. The QBM is a generative model that learns the underlying distribution of training data and generates new samples. Moreover, QGANs consist of a generator and discriminator network to create realistic data samples.
4. **Reinforcement learning:** The field of quantum computing and reinforcement learning is known as quantum reinforcement learning. It makes exploration and exploitation in decision-making processes more effective.
5. **Hybrid methods:** The variational quantum eigensolver (VQE) and quantum transfer learning algorithms fall into this category. To determine the lowest eigenvalue of a given Hamiltonian, VQE uses a hybrid quantum-classical method. Quantum transfer learning leverages pre-trained models on classical data to enhance learning performance on quantum data.
6. **Quantum neural networks (QNNs):** These utilize quantum circuits as the building blocks for various machine learning tasks.

By categorizing these algorithms, we can better understand their purpose and the specific techniques they employ. This classification provides a framework for exploring and developing quantum ML algorithms and their applications in different domains. Hence, in the remaining sections of this chapter, the working principles, features, advantages and applications of each algorithm are described.

3.3.1 Quantum Support Vector Machine (QSVM)

The quantum support vector machine (QSVM) is a hybrid algorithm that performs classification by combining classical and quantum components. It utilizes the principles of quantum computing to enhance the processing of data and improves classification accuracy. The QSVM aims to find an optimal hyperplane in a high-dimensional quantum feature space that maximally separates different classes.

The many QSVM processes are depicted in the functional schematic in Figure 3.2. The classical feature mapping function ϕ (x_i), denoted as $|\phi(x_i)\rangle$, is used to first map the classical data points (x_i) to a higher-dimensional feature space. A quantum feature map $(U\phi)$ is then used to encode the feature vectors into quantum states. The feature

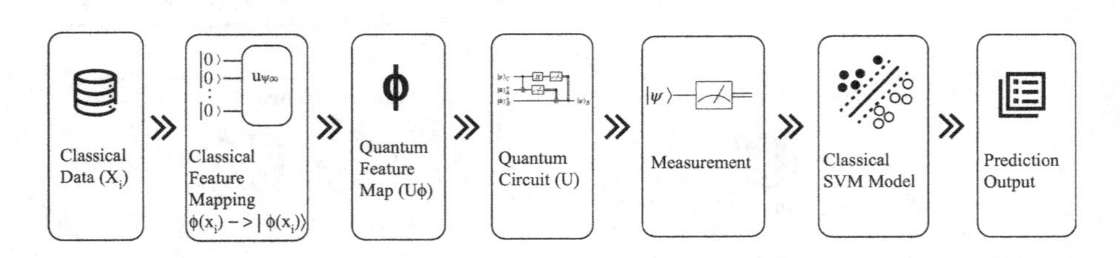

FIGURE 3.2 Functional diagram of a quantum support vector machine.

vectors are then encoded into quantum states using a quantum feature map ($U\phi$). The quantum circuit (U) is used in the encoded quantum states to perform quantum computations. Measurements are then performed on the quantum circuit to obtain classical outcomes. The classical SVM model is trained based on the measurement outcomes to classify new data points and produce prediction outputs.

To give a thorough grasp of this approach, it is important to go into the underlying arithmetic, equations, and descriptions of the QSVM [7].

3.3.1.1 Mathematical Notation and Equations

A. **Data representation:**
   - Let us consider a dataset consisting of N data points: $\{x_i, y_i\}$, where x_i represents the features of the *i*th data point and y_i denotes its corresponding class label. The class labels can be binary (+1 or −1) or multiclass encoded using one-hot encoding.

B. **Quantum feature map:**
   - A quantum feature map is used by the QSVM to translate the classical data into a quantum feature space [8]. Using a quantum circuit known as a quantum feature map, or QFM, ϕ (x), the classical data x is encoded into a quantum state $|\phi(x)\rangle$.

C. **Kernel matrix:**
   - The quantum feature vectors are used to compute the kernel matrix K.

$$K(x, x') = \langle \phi(x) | \phi(x') \rangle.$$

D. **Support vector classification:**
   - Given the kernel matrix K, the QSVM algorithm finds the support vectors, which are the data points closest to the decision border.

E. **Decision function:**
   - The decision function in QSVM is given by:

$$f(x) = \text{sign}\left(\sum_i \alpha_i y_i K(x_i, x) + b\right),$$

where b is the bias term.

3.3.1.2 Description of QSVM Algorithm

The following mathematical notations can be used to describe the QSVM algorithm.

Step 1: For each classical data point x_i:
   - Apply a quantum feature map to encode x_i into a quantum feature vector $|\phi(x_i)\rangle$.

Step 2: Compute the kernel matrix K using the quantum feature vectors $|\phi(x_i)\rangle$:

$$K(x_i, x_j) = \langle \phi(x_i) | \phi(x_j) \rangle.$$

Step 3: Find the support vectors and compute the coefficients α_i that define the decision function:

- Discover the support vectors among the x_i data points that are most near the boundaries of the decision.
- Compute the coefficients α_i using the support vectors and the class labels y_i.

Step 4: Determine the bias term b:

- Calculate the bias term b using the coefficients α_i and the class labels y_i.

Step 5: Given a new data point x, use the decision function to predict its class label:

- Compute the decision function:

$$f(x) = \text{sign}\left(\sum_i \alpha_i y_i K(x_i, x) + b\right).$$

- Based on the direction of the decision function, forecast the class label y_{pred} for the new data point x.

By applying quantum computation, the QSVM can enhance classification issues. Understanding the fundamental ideas behind the QSVM will enable researchers and practitioners to use it to tackle difficult classification problems and utilize the power of quantum computing in machine learning.

3.3.2 Quantum Neural Networks (QNN)

Traditional neural networks have revolutionized machine learning by providing powerful tools for pattern recognition and decision-making. Quantum neural networks (QNNs) aim to harness the computational capabilities of quantum systems to further enhance the learning and processing of data. The principles of quantum mechanics are applied in QNNs [9] to improve the representation and processing of data. Also, QNNs combine classical neural network architectures with quantum circuit models to create a hybrid framework. They utilize quantum states and quantum operations to represent and manipulate data, offering potential advantages in solving complex optimization problems and learning tasks. In this chapter, we explore the mathematical notation, equations, and descriptions of QNNs, highlighting their unique characteristics and potential applications.

In QNNs, quantum neurons serve as the basic processing units. They are represented by quantum states and implement quantum operations to transform the input data. Quantum neurons can be realized using quantum circuits or other quantum computational models. In a QNN, quantum layers are a collection of quantum neurons connected in a specific topology. Each neuron takes input data, performs quantum computations, and produces output data. The quantum layers are stacked to create deeper architectures, enabling the representation of more complex patterns and relationships in the data.

The functional diagram in Figure 3.3 illustrates the different steps involved in QNNs. The quantum feature map $|\phi(x)\rangle$ is used to first encode the classical input (x)

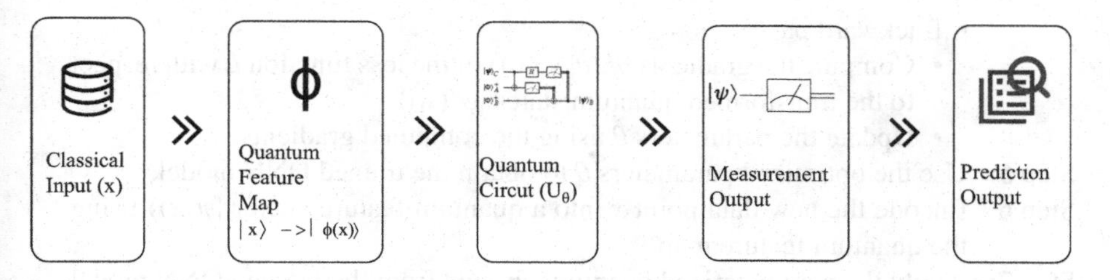

FIGURE 3.3 Functional diagram of quantum neural networks.

into the quantum state $|x\rangle$. The encoded quantum state undergoes quantum computations represented by the quantum circuit ($U\theta$), parameterized by the set of trainable parameters θ. Measurements are performed on the quantum circuit to obtain measurement outcomes, which are then used to generate classical outputs. The classical outputs represent the predictions or results of the QNN.

Typically, QNNs employ parameterized quantum circuits for training, where the training process is used to optimize the quantum gates' programmable parameters. These parameters determine the transformation applied by the quantum neurons and are learned to minimize a suitable loss function. Otherwise, QNNs are trained using another significant method called quantum backpropagation [10, 11]. It involves calculating the gradients via the parameterized quantum circuits, allowing the circuit parameters to be optimized using conventional optimization procedures.

3.3.2.1 Quantum Neural Networks (QNN) Algorithm

Step 1: For each classical data point x_i:

- Apply a quantum feature map to encode x_i into a quantum feature vector $|\psi(x_i)\rangle$.

Step 2: For each quantum layer l:

- Apply a parameterized quantum circuit $U(\theta_l)$ to the quantum feature vectors $|\psi(x_i)\rangle$.
- Obtain the transformed quantum state $|\psi'(x_i)\rangle$ as the output.

Step 3: Initialize the parameters θ for the parameterized quantum circuits.

Step 4: Iterate the following steps until convergence:

- Forward pass:
  - For each training data point (x_i, y_i):
    - To obtain the quantum feature vector, apply the quantum feature map

$$|\psi(x_i)\rangle.$$

    - Apply the quantum layers to obtain the transformed quantum state $|\psi'(x_i)\rangle$.
  - Calculate the output predictions based on the transformed quantum states

$$|\psi'(x_i)\rangle.$$

- Backward pass:
  - Compute the gradients $\partial L/\partial|\psi'(x_i)\rangle$ of the loss function L with respect to the transformed quantum states $|\psi'(x_i)\rangle$.
  - Update the parameters θ using the computed gradients.

Step 5: Use the optimized parameters θ to obtain the trained QNN model.

Step 6: Encode the new data point x into a quantum feature vector $|\psi(x)\rangle$ using the quantum feature map.

Step 7: Apply the parameterized quantum circuits from the learned QNN model to yield the converted quantum state $|\psi'(x)\rangle$.

Step 8: Make a class label prediction y_{pred} based on the transformed quantum state $|\psi'(x)\rangle$.

3.3.2.2 Applications of QNNs

- ***Quantum-enhanced pattern recognition***: By utilizing their ability for quick computing and high-dimensional data representation, QNNs can be employed for pattern recognition applications, such as picture classification or audio recognition.
- ***Optimization and combinatorial problems***: QNNs offer advantages in solving optimization and combinatorial problems by exploiting the quantum properties of superposition and entanglement. They can be used for tasks such as portfolio optimization, vehicle routing, or protein folding.
- ***Quantum generative modeling***: QNNs aid in the modeling of tasks that call for generative learning by identifying the underlying probability distribution of the input. As a result, it is possible to create fresh samples that share characteristics with the training set.

3.3.2.3 Devices to Implement QNNs

- ***Gate-based quantum computers***: Quantum neural networks (QNNs) can be implemented using gate-based quantum computers. These computers utilize quantum gates to manipulate qubits and perform quantum operations. These computers which are based on the circuit model allow for the execution of parameterized quantum circuits and enable the training and inference processes. To implement QNNs on gate-based quantum computers, one needs to design quantum circuits that represent the neural network architecture. This entails mapping the layers, neurons, and connections of the neural network onto the quantum computer's qubits and gates. The quantum gates are then applied to perform computations and update the quantum state of the qubits.

  However, gate-based quantum computers currently face challenges such as limited qubit coherence times, high error rates, and the need for error correction. These limitations can impact the performance and scalability of QNNs on gate-based quantum computers. Nonetheless, ongoing research and advancements in quantum computing technology aim to address these challenges and improve the feasibility of using gate-based quantum computers for QNNs.

- ***Variational quantum circuits (VQC)*:** Variational quantum circuits (VQCs) are a promising approach for implementing QNNs as they combine classical and quantum computations to train and optimize the parameters of a quantum circuit. In the context of QNNs, VQCs can be used to represent the neural network architecture and perform computations on quantum states. Using classical optimization procedures, the quantum circuit's parameters are modified iteratively in order to reduce a cost function that is commonly connected to the QNN's training goal.

  VQCs offer several advantages for QNNs. They can be implemented on existing gate-based quantum computers or simulators, making them accessible for practical applications. Additionally, VQCs can leverage the power of quantum parallelism and potentially provide computational advantages over classical neural networks in certain scenarios. However, it's important to note that VQCs also face challenges. The optimization process can be computationally intensive, requiring a large number of iterations to find optimal parameter values. Furthermore, the performance of VQCs can be affected by noise and errors in the quantum hardware. Despite these challenges, VQCs hold promise for QNNs and are an active area of research. Ongoing efforts aim to improve the performance and scalability of VQCs, making them a viable option for implementing QNNs on quantum computing platforms.

Other devices that can be used to implement QNNs are quantum annealers, photonic quantum computers and quantum simulators. Besides, there are several quantum computing platforms available for implementing QNNs. Some popular platforms include IBM Quantum, Google Quantum Computing, Microsoft Quantum Development Kit, and Rigetti Forest. These platforms provide access to quantum hardware and software tools for developing and running QNNs.

3.3.3 Variational Quantum Eigensolver (VQE)

For addressing issues in quantum chemistry and optimization, a variational quantum eigensolver (VQE) is recognized as an effective quantum machine learning technique [12]. Calculation of the energy of the ground state of a Hamiltonian is performed using a quantum method known as the VQE. It is especially beneficial for addressing issues in quantum chemistry and materials science, and is a promising method for tackling issues that are challenging for conventional computers to solve because of their exponential complexity. But it's important to remember that VQE's performance is now constrained by the noise and error rates of the existing quantum hardware.

When utilizing the VQE algorithm to determine the ground state energy, the Hamiltonian, which in VQE stands in for the overall energy of a quantum system, is extremely important. In VQE, the Hamiltonian is typically decomposed into two parts: the classical part, which represents the system's classical energy, and the quantum part, which represents the system's quantum energy. The quantum part is usually expressed as a sum of terms, each corresponding to a specific interaction or property of the system.

The functional diagram in Figure 3.4 illustrates the different steps involved in VQE. The quantum circuit (U) represents the parameterized quantum circuit used to

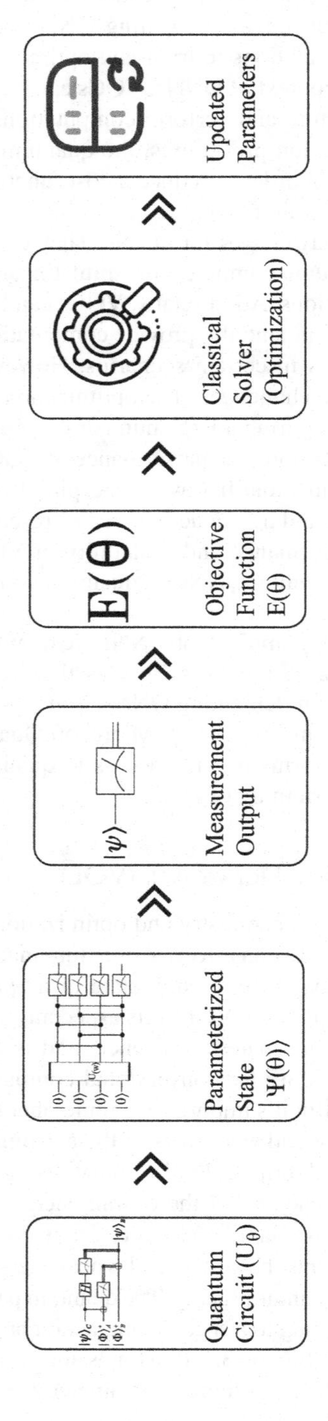

FIGURE 3.4 Functional diagram of a variational quantum eigensolver.

prepare a state $|\psi(\theta)\rangle$, where θ represents the set of trainable parameters. Measurements are performed on the state to obtain measurement outcomes. The results of these measurements are used to assess an objective function $E(\theta)$, which commonly denotes the Hamiltonian's expectation value for the targeted quantum system. After that, a traditional solver or optimization technique is used to minimize the objective function. The quantum circuit's parameters (θ) are iteratively updated during the optimization process until the desired result is attained.

3.3.3.1 Mathematical Notation and Equations

A. **Problem formulation:**
- Consider a quantum system that is described by the Hamiltonian operator H. The VQE technique seeks to identify the H eigenvector with the lowest eigenvalue (ground state energy), which may provide crucial information about the system's properties and behavior.

B. **Variational principle:**
- The expectation value of the Hamiltonian operator H is an upper constraint on the system's ground state energy for every trial wavefunction $\Psi(\theta)$ depending on a set of parameters θ, according to the variational principle, which is used by the VQE method.

$$E(\theta) = \langle \Psi(\theta) | H | \Psi(\theta) \rangle \geq E_0,$$

where E_0 is the system's actual ground state energy.

C. **Parameterized quantum circuit:**
- The trial wavefunction $\Psi(\theta)$ is prepared by the VQE method using a parameterized quantum circuit represented by $U(\theta)$. This quantum circuit consists of a sequence of quantum gates acting on the initial state, typically the all-zero state $|0\rangle$.

D. **Energy estimation:**
- To estimate the expectation value $E(\theta) = \langle \Psi(\theta)|H|\Psi(\theta)\rangle$, the VQE algorithm uses quantum measurements on the quantum state prepared by $U(\theta)$. By performing measurements of H on the prepared state, the energy expectation value can be estimated.

E. **Optimization loop:**
- The VQE approach uses an optimization loop to identify the set of parameters that minimizes the energy expectation value. This optimization procedure frequently makes use of classical optimization techniques like gradient descent or Bayesian optimization to update the parameters and decrease energy.

F. **Convergence criteria:**
- The convergence of the VQE algorithm is determined by a predefined convergence criterion. This criterion can be based on the energy difference between consecutive iterations or the convergence of the parameter updates.

G. **Application: Quantum chemistry:**
- VQE has essential uses in quantum chemistry, where it may be applied to calculate the energy of a system's ground state. By encoding the molecular Hamiltonian into the VQE framework, one can obtain insights into chemical properties, bond lengths, reaction rates, and more.

H. **Extension: Hybrid classical-quantum optimization:**
- In some cases, the VQE algorithm can be combined with classical optimization techniques to solve larger-scale problems. This hybrid approach leverages the strengths of classical optimization algorithms while benefiting from the quantum computational advantages of the VQE algorithm.

3.3.3.2 Variational Quantum Eigensolver (VQE) Algorithm

The VQE algorithm can be described as follows:

Input: Hamiltonian operator H, parameterized quantum circuit $U(\theta)$, classical optimization algorithm, initial parameters θ_{init}, convergence threshold ε.

Output: Optimal parameters θ_{opt}, corresponding minimum eigenvalue E_{min}.

Algorithm:

Step 1: Initialize the parameters: Set $\theta \leftarrow \theta_{\text{init}}$.

Step 2: Perform the following steps iteratively until convergence:

a. Prepare the trial wavefunction: Apply the parameterized quantum circuit $U(\theta)$ to an initial quantum state, typically the all-zero state $|0\rangle$, to prepare the trial wavefunction $|\Psi(\theta)\rangle$.

b. Evaluate the energy: Measure the expectation value of the Hamiltonian operator H with respect to the trial wavefunction:

$$E_{\text{current}} \leftarrow \langle \psi(\theta) | H | \psi(\theta) \rangle.$$

c. Update the parameters: Using the energy estimation E_{current}, update the parameters using a standard optimization process. This optimization aims to minimize the energy expectation value.

d. Check convergence: The iteration should end, and the next step should be taken if the difference between the current energy E_{current} and the prior energy E_{previous} is less than the convergence threshold ε.

e. Update the previous energy: Set $E_{\text{previous}} \leftarrow E_{\text{current}}$.

Step 3: Output the results: Return the optimal parameters θ_{opt} and the corresponding minimum eigenvalue E_{min} obtained after convergence.

Note: The classical optimization algorithm used in step 2c can be any suitable method, such as gradient descent, Nelder-Mead, or Bayesian optimization, depending on the problem and available resources. The convergence threshold ε determines the desired level of precision in the energy estimation.

Hence, by applying the above algorithm, the minimal eigenvalue and accompanying eigenvector of a given Hamiltonian operator can be estimated. This method not only uses a parameterized quantum circuit and classical optimization to identify the ideal parameters that minimize the energy expectation value but also gets close to the quantum system's ground state energy by iteratively adjusting the parameters.

3.3.4 Quantum Boltzmann Machine (QBM)

The classical Boltzmann machines served as the inspiration for the quantum Boltzmann machine (QBM); a form of quantum neural network [13]. Quantum mechanics (QM) is a methodology for doing probabilistic modeling and learning problems and QBMs are networks of connected nodes, or qubits in the case of quantum computing, similar to classical Boltzmann machines. The exploration of complex probability distributions is made possible by the interaction of these qubits with one another through quantum gates.

The QBM utilizes quantum effects such as superposition and entanglement to enhance its computational capabilities. By exploiting these quantum properties, QBM can potentially offer advantages over classical Boltzmann Machines in terms of computational power and efficiency.

The functional diagram Figure 3.5 illustrates the key steps involved in the QBM algorithm. The classical data points (x_i) are initially provided as input. These classical data points are then encoded into quantum feature vectors using a quantum feature map, represented as $\phi(x_i)$. The quantum feature vectors are used to compute the quantum kernel matrix, denoted as $K(x_i, x_j)$, which captures the similarity between quantum feature vectors. From the quantum kernel matrix, a quantum Boltzmann distribution $p(x)$ is generated. Samples are then drawn from this distribution, denoted as x_i', which are used to update the classical data points through a classical data update step ($x_i \leftarrow x_i'$). This iterative process continues until convergence, or a desired stopping criterion is met.

3.3.4.1 Characteristics of Quantum Boltzmann Machine (QBM)

The classical Boltzmann machine (CBM) serves as the inspiration for the QBM. The CBM is a stochastic generative model consisting of binary units, also known as neurons, connected through weighted connections. It learns the underlying distribution of the training data by adjusting the weights through a process called Gibbs sampling.

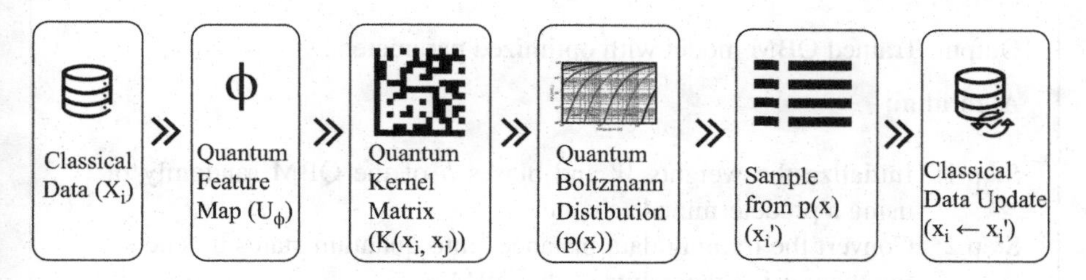

FIGURE 3.5 Functional diagram of quantum Boltzmann machine.

The QBM extends the concept of the CBM to the realm of quantum computing. It replaces the binary units in the CBM with quantum bits, or qubits, and the weighted connections with quantum gates. The QBM leverages quantum superposition and entanglement to perform computations that classical computers find challenging.

- In the QBM, the compatibility between the state of the qubits and the training data is represented by an energy function. It is defined by a Hamiltonian operator, denoted as H. The Hamiltonian encodes both the data and the interaction terms between qubits.
- The QBM utilizes the Gibbs state, also known as the thermal equilibrium state, to model the probability distribution of the qubits.
- The partition function, denoted as Z, normalizes the probabilities and is crucial for calculating the expectation values.
- The training of the QBM entails identifying the ideal weights and biases that reduce the energy of the qubits. Usually, methods like variational optimization or gradient descent are used to do this. The objective is to discover a training set representation that captures the training set's statistical properties.
- The QBM can be utilized for generative modeling, thereby, allowing the generation of new samples from the learned distribution. By sampling from the QBM's quantum state, new data points can be generated that resemble the training data.
- The QBM supports unsupervised learning tasks like dimensionality reduction and clustering. The latent characteristics and patterns that help with data analysis and representation may be captured by QBM by understanding the underlying distribution of the data.
- The QBM can be used for optimization problems, which takes advantage of quantum annealing techniques. By mapping the optimization problem onto the QBM's energy landscape, quantum annealing can be leveraged to search for the optimal solution efficiently.

3.3.4.2 Quantum Boltzmann Machine (QBM) Algorithm

Input: Training dataset D number of qubits N, learning rate α, number of iterations T.

Output: Trained QBM model with optimized parameters.

Algorithm:

Step 1: Initialize the weights W and biases b of the QBM randomly or using a predetermined scheme.

Step 2: Convert the training data instances into quantum states by encoding them using the qubits of the QBM.

Step 3: Define the Hamiltonian operator H that represents the energy function of the QBM. The Hamiltonian contains terms that capture the compatibility between qubits and the training data.

Step 4: Prepare the quantum state of the QBM by setting the qubits to an initial state, such as the equal superposition state.

Step 5: Perform Gibbs sampling:

For each iteration t from 1 to T, perform the following steps:

a. Calculate the energy E of the quantum state by measuring the expectation value of the Hamiltonian H.

$$E = \langle \Psi | H | \Psi \rangle$$

b. Update the quantum state using a quantum circuit, such as a series of quantum gates, based on the current values of W and b.

c. Measure the values of the qubits in the quantum state to obtain a sample from the current distribution.

d. Update the QBM parameters W and b using a learning rule, such as gradient descent or stochastic gradient descent, to minimize the energy of the QBM. The updates can be performed as:

$$W \leftarrow W - \alpha \partial E / \partial W$$

$$b \leftarrow b - \alpha \partial E / \partial b$$

Step 6: Return the QBM model with the optimized parameters W and b.

Note: The specific form of the energy function E, the update rule for the QBM parameters in step 5d, and the choice of quantum gates depend on the specific implementation and optimization approach.

The QBM algorithm iteratively optimizes the QBM parameters through Gibbs sampling, quantum state updates, and parameter updates. By minimizing the energy of the QBM, it learns the underlying distribution of the training data. After training, the QBM can be employed in a variety of tasks, including generative modeling, unsupervised learning, and optimization.

3.3.5 Quantum Principal Component Analysis (QPCA)

Quantum machine learning algorithms have gained significant attention in recent years for their potential to outperform classical machine learning techniques in certain tasks. One such algorithm is quantum principal component analysis (QPCA) [14], which combines the principles of quantum computing and the mathematical framework of principal component analysis (PCA). Quantum principal component

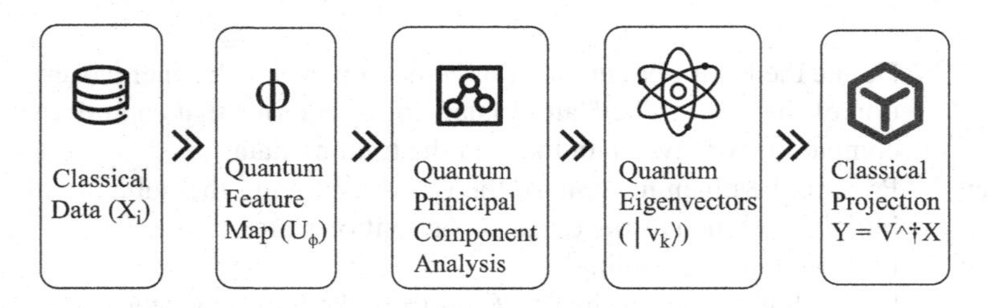

FIGURE 3.6 Functional diagram of quantum principal component analysis.

analysis aims to extract the most informative features or principal components from a dataset, enabling efficient dimensionality reduction and data analysis.

The functional diagram Figure 3.6 illustrates the key steps involved QPCA using mathematical notations. The classical data matrix X is provided as input. The classical data points x_i are then encoded into quantum feature vectors $|\phi(x_i)\rangle$ using a quantum feature map. The QPCA algorithm is applied to the quantum feature vectors to extract the quantum eigenvectors ($|vk\rangle$), which represent the principal components. These quantum eigenvectors are then used to perform a classical projection step, where the classical data matrix Y is obtained by projecting X onto the quantum eigenvectors ($Y = V†X$). The resulting classical data matrix Y represents the transformed data in the QPCA space.

3.3.5.1 QPCA—Mathematical Notation and Preliminaries

To understand QPCA, the following mathematical notations are established:

- Let X be the classical dataset with d-dimensional feature vectors: $X = \{x_1, x_2, x_3 \ldots, x_n\}$, where each $x_i \in \mathbb{R}^d$.
- Let $|\phi(x)\rangle$ denote the quantum state corresponding to a classical data point x.
- Let $U(x)$ be the quantum feature map that encodes a classical data point x into a quantum state: $|\phi(x)\rangle = U(x)|0\rangle$.
- Let $|\Psi\rangle$ denote the quantum state corresponding to the entire dataset: $|\Psi\rangle = \Sigma_i|\phi(x_i)\rangle$.

A. **Quantum feature map**
   - A key component of QPCA is the quantum feature map. It utilizes a quantum circuit to convert classical data points into quantum states. This transformation can be represented as:

$$|\phi(x)\rangle = U(x)|0\rangle,$$

where $|0\rangle$ represents the initial state and $U(x)$ is the unitary transformation corresponding to the quantum feature map. The selection of the quantum feature map relies on the specific problem and can range from simple rotations to more complex circuits.

B. **Quantum singular value estimation**
   - The QPCA algorithm involves estimating the singular values of the quantum feature matrix. The quantum feature matrix is constructed by applying the quantum feature map to the classical dataset. The singular values can be estimated using quantum algorithms and techniques, such as quantum phase estimation. These estimates provide insights into the importance of each principal component and guide the dimensionality reduction process.

C. **Principal component selection**
   - Once the singular values are estimated, the next step is to select the top principal components based on their corresponding singular values. The dataset's maximum variance's directions are captured by the principal components, which also offer a compressed representation of the data. The selection of the top principal components can be done using various methods, such as thresholding or retaining a certain percentage of the total variance.

D. **Principal component reconstruction**
   - Using the chosen primary components, the dataset may be rebuilt using the QPCA technique. The original dataset can be projected onto the chosen main components to create the rebuilt dataset. The process of reconstruction can be described as:

$$X' = \Sigma_j (\sigma_j \mid u_j \rangle \langle u_j \mid) X,$$

where X' is the reconstructed dataset, σ_j are the singular values corresponding to the selected principal components, and $|u_j\rangle$ are the corresponding eigenvectors.

3.3.5.2 Algorithm for Quantum Principal Component Analysis

Input: Classical dataset $X = \{x_1, x_2, x_3 ..., x_n\}$, number of principal components k.

Output: Selected principal components $U = \{u_1, u_2, ..., u_k\}$.

Algorithm:

Step 1: Quantum feature map:
- For each classical data point $x_i \in X$, apply a quantum feature map $U(x_i)$ to encode it into a quantum state: $|\phi(x_i)\rangle = U(x_i)|0\rangle$.

Step 2: Quantum singular value estimation:
- Apply quantum phase estimation or other quantum algorithms to estimate the singular values $\sigma_1, \sigma_2, ..., \sigma_d$ of the quantum feature matrix.
- Sort the singular values in descending order.

Step 3: Principal component selection:
- Select the top-k singular values and their corresponding eigenvectors.
- Set $U = \{u_1, u_2, ..., u_k\}$ as the selected principal components.

Step 4: Principal component reconstruction:

- For each classical data point $x_i \in X$, calculate its reconstructed value $x'i$ using the selected principal components:

$$x'i = \Sigma_j \left(\sigma_j u_j\right)\left(u_j\right)^{\mathrm{T}} x_i.$$

Step 5: Return U as the set of selected principal components.

This algorithm outlines the steps involved in performing QPCA on a classical dataset. It starts by encoding the classical data points into quantum states using a quantum feature map. Then, the singular values of the quantum feature matrix are estimated using quantum algorithms. The top-k singular values and their corresponding eigenvectors are selected as the principal components. Finally, the dataset is reconstructed using the selected principal components.

Quantum principal component analysis (QPCA) combines the power of quantum computing and the principles of principal component analysis (PCA) to extract essential features from classical datasets. By leveraging quantum properties, QPCA offers potential advantages in terms of computational efficiency and information extraction. It offers efficient dimensionality reduction, which is beneficial for tasks such as data visualization, clustering and classification. Additionally, QPCA can also be utilized for feature selection and extraction in quantum data analysis and has the potential to provide insights into quantum states and quantum dynamics.

3.3.6 Quantum K-Means Clustering (QK-Means)

Numerous applications, including data analysis, pattern identification, and optimization, show considerable promise for quantum K-means clustering (QK-Means) [15]. Compared to classical approaches, it can produce clustering results that are more effective and precise because it can take advantage of quantum superposition and quantum parallelism.

The functional diagram in Figure 3.7 illustrates the key steps involved in QK-Means clustering using mathematical notations. The classical data matrix X is provided as input. The classical data points x_i are then encoded into quantum feature

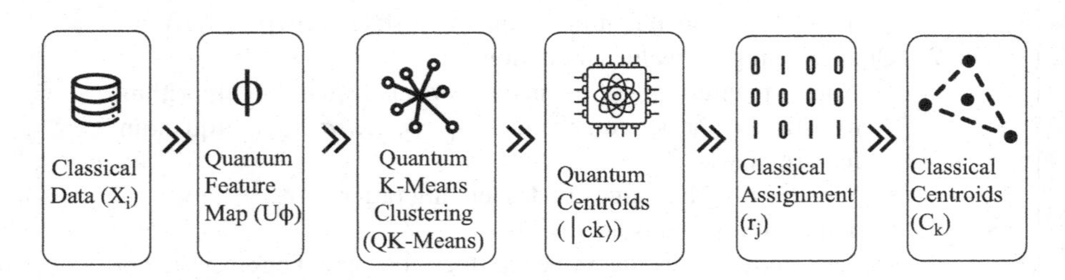

FIGURE 3.7 Functional diagram of quantum K-means clustering.

vectors $|\phi(x_i)\rangle$ using a quantum feature map. The QK-Means algorithm is applied to the quantum feature vectors to determine the quantum centroids $|ck\rangle$. A classical assignment step is performed to assign each data point to the nearest quantum centroid, resulting in the classical assignment vector r_i. Finally, the classical centroids ck are computed based on the assigned data points. The resulting classical centroids represent the final clusters obtained through QK-Means clustering.

3.3.6.1 Algorithm of Quantum K-Means Clustering

Quantum K-means clustering is a quantum machine learning algorithm that aims to partition a dataset into k clusters. It is an extension of the classical K-means clustering algorithm that utilizes quantum techniques to potentially provide improved clustering results. The algorithm steps are presented as below:

Input: Classical dataset $X = \{x_1, x_2, \ldots, x_n\}$, number of clusters k.

Output: Cluster centroids $C = \{c_1, c_2, \ldots, c_k\}$.

Algorithm:

Step 1: Quantum data encoding:
- For each classical data point $x_i \in X$, apply a quantum feature map $U(x_i)$ to encode it into a quantum state: $|\phi(x_i)\rangle = U(x_i)|0\rangle$.

Step 2: Initial centroid initialization:
- Randomly select k quantum states $|c_j\rangle$ as initial centroids.

Step 3: Quantum distance calculation:
- For each data point $|\phi(x_i)\rangle$, calculate the quantum distance to each centroid $|c_j\rangle$:

$$D(x_i, c_j) = \langle \phi(x_i) | M_j | \phi(x_i) \rangle,$$

where M_j is a Hermitian operator representing the centroid $|c_j\rangle$.

Step 4: Quantum distance-based assignment:
- Assign each data point $|\phi(x_i)\rangle$ to the closest centroid $|c_j\rangle$ based on the calculated quantum distances.

Step 5: Centroid update:
- For each cluster, calculate the updated centroid by taking the average of the data points assigned to that cluster:

$$|c_j\rangle = (1/|S_j|)\Sigma_i \in S_j |\phi(x_i)\rangle,$$

where S_j is the set of data points assigned to cluster j.

Step 6: Repeat steps 3–5 until convergence or maximum iterations reached.

Step 7: Return $C = \{c_1, c_2, \ldots, c_k\}$ as the final cluster centroids.

This algorithm lays out how to execute quantum K-means clustering. It starts by employing a quantum feature map to encode the classical data points into quantum states. From the quantum states, the initial centroids are chosen at random. The quantum distances between the centroids and data points are then determined. Based on the quantum distances, data points are assigned to the nearest centroid. By averaging the assigned data points, the centroids are updated. The final cluster centroids are returned after repeating these stages until convergence or the maximum number of iterations is reached.

3.3.6.2 Challenges in Quantum K-Means Clustering (QK-Means)

Quantum K-means clustering offers a quantum-inspired approach to data clustering, leveraging the computational power of quantum systems. As quantum hardware continues to advance, QK-Means clustering algorithms are expected to play a vital role in various domains requiring advanced data analysis and pattern recognition. However, there are several challenges and considerations when implementing QK-Means clustering. These include:

A. **Quantum hardware limitations:** Quantum computers are still in their early stages of development, with limited qubit counts, noise, and error rates. These limitations impact the scalability and accuracy of QK-Means clustering algorithms.
B. **Quantum feature map design:** The selection of an appropriate quantum feature map, $\phi(x)$, is crucial for successful clustering. The choice of the feature map can significantly impact the representation and separability of the data in the quantum state space.
C. **Distance metric operator:** Designing an effective distance metric operator, M_j, is essential for computing the quantum distance between data points and centroids accurately. The choice of the distance metric affects and determines the clustering performance.
D. **Quantum circuit depth:** The depth of the quantum circuit used for encoding and distance calculation can impact the algorithm's runtime and the coherence of the quantum states. Minimizing the circuit depth is crucial for mitigating errors and maintaining quantum coherence.

Despite these challenges, QK-Means clustering shows promise as a quantum machine learning algorithm. Researchers are actively exploring novel techniques, such as variational quantum algorithms and hybrid classical-quantum approaches, to improve its performance and scalability.

3.3.7 Quantum Generative Adversarial Networks (QGANs)

The QGANs are a powerful framework that combines concepts from generative modeling and adversarial learning within the quantum computing domain [16]. Quantum simulation, quantum chemistry, and quantum machine learning might all benefit from using QGANs to produce realistic quantum data distributions. A generator network and a discriminator network are the two primary parts of the deep learning

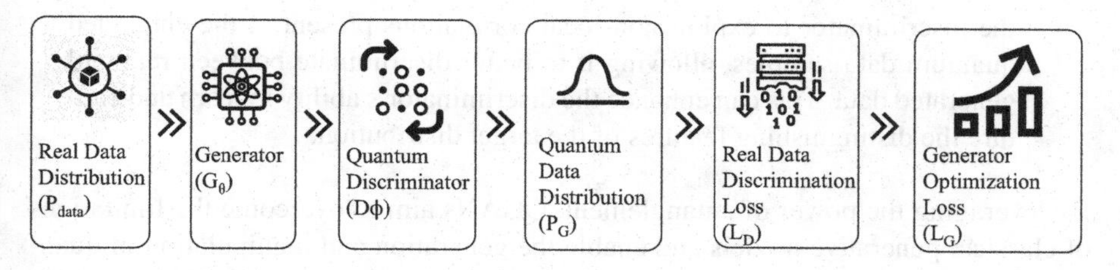

FIGURE 3.8 Functional diagram of QGANs.

model known as generative adversarial networks (GANs). A training dataset will be used to create new data samples for GANs to analyze. The generator and discriminator networks are trained in a competitive way according to the architecture of GANs, which is based on an adversarial framework.

The functional diagram in Figure 3.8 represents the key components of quantum generative adversarial networks (QGANs) using mathematical notations. The real data distribution P_{data} serves as the reference for generating realistic data. The generator G_θ produces quantum data samples that mimic the real data distribution. The quantum discriminator $D\phi$ evaluates the generated data distribution P_{G} and distinguishes between real and generated data. The discriminator loss L_{D} measures the discrepancy between the discriminator's predictions and the true labels. The generator loss L_{G} quantifies the generator's ability to deceive the discriminator. The optimization process aims to minimize both L_{D} and L_{G}, enabling the generator to generate realistic data samples that approximate the real data distribution.

3.3.7.1 Architecture of Quantum Generative Adversarial Networks (QGANs)

The QGAN architecture combines classical and quantum components to generate and learn quantum data distributions. By using quantum resources and techniques, QGANs aim to overcome the limitations of classical generative models and enable the generation and analysis of quantum data in various quantum applications, such as quantum machine learning and quantum simulation.

In the context of QGANs, quantum entanglement plays a crucial role in generating and learning quantum data distributions. The relevance of quantum entanglement in QGANs can be understood in two key aspects:

A. **Quantum data generation:** In a QGAN, the generator network is designed to generate quantum data samples that resemble a target quantum distribution. Quantum entanglement allows the generator to create entangled states, which can exhibit complex correlations and coherence properties that are characteristic of the target distribution. By leveraging entanglement, QGANs aim to generate quantum states that capture the statistical properties of the target distribution more accurately.
B. **Quantum discrimination:** The discriminator network in QGANs is responsible for distinguishing between real quantum data samples and the synthetic data samples generated by the generator. Quantum entanglement enables

the discriminator to exploit non-local correlations present in the entangled quantum data samples, allowing it to better discriminate between real and generated data. This can enhance the discriminator's ability to learn and capture the distinguishing features of the target distribution.

By leveraging the power of entanglement, QGANs aim to overcome the limitations of classical generative models and enable the generation and manipulation of quantum data. Quantum entanglement provides a unique resource for QGANs to generate realistic quantum states and capture the statistical properties of quantum distributions, which is crucial the system's actual ground state energy.

A high-level overview of the QGAN architecture is described as follows:

A. **Quantum generator:** The quantum generator is responsible for generating quantum data samples that mimic a target quantum distribution. It takes as input a set of quantum parameters, usually represented as a set of quantum gates or circuits. The generator applies these quantum operations to an initial state, such as a quantum register or a quantum circuit, and generates a quantum state that represents a synthetic data sample. The generator aims to produce quantum states that resemble the statistical properties of the target distribution.
B. **Quantum discriminator:** The quantum discriminator's role is to differentiate between real quantum data samples from the target distribution and the synthetic data samples generated by the generator. It takes as input quantum states and performs measurements or other quantum operations on them. The discriminator aims to accurately classify whether the input quantum state is real or generated by the generator.
C. **Quantum training:** Similar to classical GANs, QGANs use an adversarial training process to optimize the generator and discriminator. Iterative adversarial training is done on the generator and discriminator. The training process alternates between two phases:
   - ***Generator training*:** The generator takes quantum parameters and generates synthetic quantum data samples. The discriminator then receives these generated samples and provides feedback on their authenticity. The generator aims to adjust its parameters to generate quantum states that fool the discriminator into classifying them as real.
   - ***Discriminator training*:** The discriminator takes both real quantum data samples from the target distribution and the generated samples from the generator. It tries to correctly classify the input quantum states as real or generated by optimizing its own quantum operations or measurements.
D. **Quantum loss functions:** Quantum loss functions are used by QGANs to direct the training of the discriminator and generator. The objective of these loss functions is normally to measure the disparity or difference between the statistical characteristics of genuine quantum data samples and synthetic quantum data samples. Fidelity-based measures, quantum divergence measures, or quantum distinguishability measures are a few examples of quantum loss functions.

E. **Quantum resources:** To create and analyze quantum data, QGANs make use of quantum resources like quantum gates, quantum circuits, and quantum measurements. The particular implementation and the properties of the desired quantum data distribution determine the choice of quantum resources. With the use of these resources, the QGAN is able to create and control quantum states by taking advantage of the special qualities of quantum systems.

3.3.7.2 Algorithm of QGANs

Input: Number of quantum data qubits N, classical data dimension D, number of generator training iterations T, learning rate α.

Output: Trained QGAN model with optimized generator and discriminator parameters.

Algorithm:

Step 1: Initialize QGAN parameters:

- Initialize the generator parameters θg randomly or using a predetermined scheme.
- Initialize the discriminator parameters θd randomly or using a predetermined scheme.

Step 2: Quantum generator function:

- Define the quantum generator function $G(\theta g)$ that takes as input a quantum state $|0\rangle$^$\otimes N$ and the generator parameters θg, and applies a quantum circuit to generate a quantum state $|\psi g(\theta g)\rangle$.

Step 3: Quantum discriminator function:

- Define the quantum discriminator function $D(\theta d)$ that takes as input a quantum state $|\psi\rangle$ and the discriminator parameters θd, and applies a quantum circuit to perform a measurement to distinguish between real and generated samples.

Step 4: Loss functions:

- Define the quantum generator loss function $L_G(\theta g, \theta d)$ that quantifies the difference between the discriminator's classification of the generated samples and the desired outcome.
- Define the quantum discriminator loss function $L_D(\theta d, \theta g)$ that measures the discriminator's ability to correctly classify real and generated samples.

Step 5: Training loop:

- For iteration t from 1 to T do:
  - Generate a batch of D-dimensional classical random vectors x from a known distribution.
  - Using the quantum generator $G(\theta g)$, encode the classical data into the quantum states $|\psi x\rangle$.

- Measure the quantum states using the quantum discriminator $D(\theta d)$ to obtain the discriminator's predictions for real and generated samples.
- Compute the quantum generator loss $L_G(\theta g, \theta d)$ using the discriminator predictions and desired outcomes.
- Compute the quantum discriminator loss $L_D(\theta d, \theta g)$ using the discriminator predictions and true labels.
- Update the generator parameters θg using gradient descent: $\theta g \leftarrow \theta g - \alpha * \nabla\theta g L_G(\theta g, \theta d)$.
- Update the discriminator parameters θd using gradient descent: $\theta d \leftarrow \theta d - \alpha * \nabla\theta d$ L_D($\theta d, \theta g$).

Step 6: Output:

- Return the trained QGAN model with the optimized generator parameters θg and discriminator parameters θd.

3.3.8 Quantum Transfer Learning (QTL)

In traditional machine learning, transfer learning is a potent technique that enables models developed for one job to be modified and applied to related tasks. Nevertheless, the introduction of quantum computing presents a chance to investigate how quantum capabilities might be used for transfer learning. This chapter explores the idea of quantum transfer learning, which tries to improve knowledge transfer between tasks by taking advantage of quantum features.

The functional diagram in Figure 3.9 represents the key components of quantum transfer learning using mathematical notations. The source task involves a pre-trained quantum model *Ms*, which captures knowledge from a previous task. The classical knowledge transfer process *T* facilitates the transfer of relevant information from the source task to the target task. The target task requires a quantum model *Mt*, which is initialized with the transferred knowledge. Fine-tuning *F* is performed to adapt the transferred model to the target task by adjusting its parameters. The final result is the transferred quantum model *Mt′*, which has been fine-tuned and optimized for the target task.

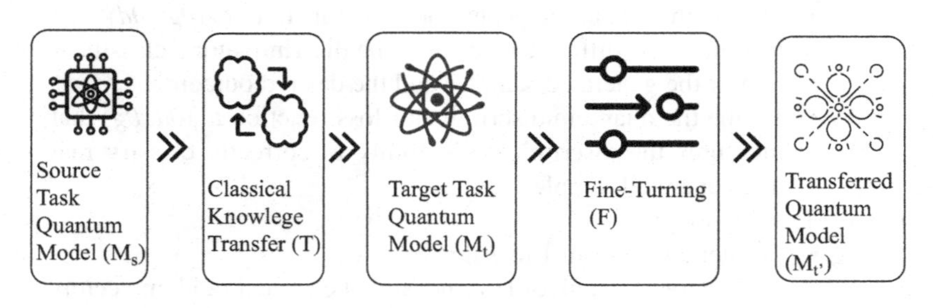

FIGURE 3.9 Functional diagram of quantum transfer learning.

3.3.8.1 Quantum Transfer Learning Framework

A framework called quantum transfer learning (QTL) makes use of previously learned quantum models and knowledge transfer strategies to facilitate learning and performance of novel quantum tasks. [17]. Here's a description of the key components in a QTL framework:

A. **Pre-trained quantum model:** The QTL framework starts with a pre-trained quantum model that has been trained on a source task or dataset. This pre-trained model serves as a knowledge source and provides a foundation of learned quantum features and parameters.
B. **Target quantum task:** The target quantum task is the new task or dataset for which the QTL framework aims to improve performance. It could be a different but related task to the source task, or it could involve a different quantum data representation or encoding.
C. **Feature extraction:** In the QTL framework, feature extraction is performed using the pre-trained quantum model. The pre-trained model is used to extract relevant quantum features or representations from the target quantum data. These qualities serve to better the learning process for the target task by capturing the knowledge acquired from the source task.
D. **Transfer learning:** To transfer the knowledge acquired from the source task to the target task, transfer learning techniques are used. This involves adapting the pre-trained quantum model or its features to the target task. Various transfer learning strategies can be employed, such as fine-tuning the model's parameters, freezing certain layers, or learning task-specific layers while preserving the shared features.
E. **Task-specific learning:** After the transfer learning phase, the QTL framework proceeds with task-specific learning on the target quantum task. This typically involves further training the adapted model or fine-tuning the learned features to optimize performance on the target task. The model's parameters may be adjusted using additional quantum data from the target task, which will increase the model's capacity to complete the task.
F. **Evaluation and performance:** The performance of the QTL framework is assessed by evaluating the adapted model on the target task. The evaluation metrics can vary depending on the specific target task, such as accuracy, error rates, or other task-specific performance measures.

Overall, the QTL framework enhances learning effectiveness and performance on the target task by allowing the transfer of knowledge and features from a pre-trained quantum model to a new target quantum job. The nature of the source and target tasks, as well as the selected transfer learning methodologies, might influence the specific implementation and architecture of the QTL framework.

3.3.8.2 Algorithm of Quantum Transfer Learning (QTL)

A general outline of the steps involved in a typical QTL process is discussed below.

Input: Source task data (Xs, Ys), Target task data (Xt, Yt)

Output: Adapted model parameters for the target task (θt)

Algorithm:

Step 1: Quantum pre-training:
- Initialize the source model parameters: θs
- Encode the source task data into quantum states:
  - For each data point xs in Xs:
  - Encode xs as $|\phi(xs)\rangle$
- Train the source model using the encoded quantum states and source task labels:
  - Use a quantum learning algorithm (e.g., quantum variational circuits) to optimize the source model parameters θs

Step 2: Quantum Fine-tuning:
- Initialize the target model parameters: θt
- Encode the target task data into quantum states:
  - For each data point xt in Xt:
  - Encode xt as $|\phi(xt)\rangle$
- Transfer information from the source model's prior training to the target model:
  - Copy the source model parameters θs to the target model parameters θt
- Train the target model using the encoded quantum states and target task labels:
  - To optimize the target model parameters θt, use a hybrid quantum-classical optimization technique.
  - Update θt based on the target task data and the transferred knowledge $\Psi(\theta s)$ from the source model: $\theta t \leftarrow \theta t - \alpha \nabla L(\theta t, Yt, \Psi(\theta s))$ where α is the learning rate and L is the loss function that compares the model predictions to the target task labels.

Step 3: Output
- Return the adapted model parameters for the target task: θt

It is important to note that the specific details and algorithms used in each step may differ depending on the research paper or implementation.

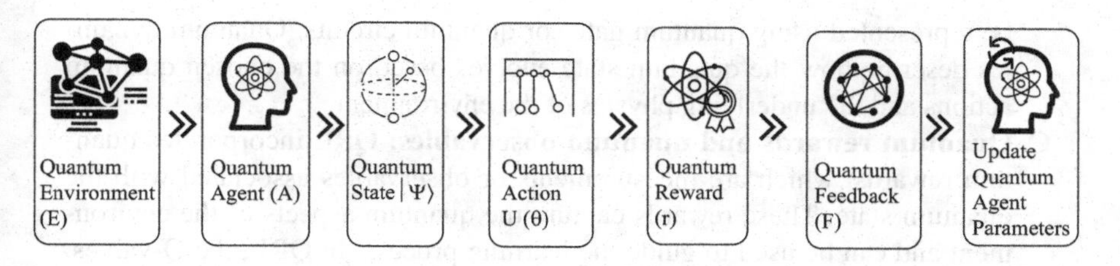

FIGURE 3.10 Functional diagram of quantum reinforcement learning.

3.3.9 Quantum Reinforcement Learning (QRL)

Quantum reinforcement learning (QRL) combines the principles of reinforcement learning and quantum computing to address complex decision-making problems [18]. By leveraging the unique properties of quantum systems, QRL aims to enhance the efficiency and scalability of reinforcement learning algorithms. In this section, we explore the foundations of QRL, its mathematical notation, and the algorithmic framework for quantum-enhanced reinforcement learning.

The functional diagram in Figure 3.10 represents the key components of QRL using mathematical notations. The quantum environment E interacts with the quantum agent A, which takes actions represented by the unitary operator $U(\theta)$ based on the current quantum state $|\psi\rangle$. The resulting quantum reward r is obtained from the environment as feedback. This feedback is used to update the parameters of the quantum agent, optimizing its behavior. This update process is performed through quantum operations and classical computations. The loop continues as the quantum agent interacts with the environment, learns from the rewards, and updates its parameters to improve its performance over time.

3.3.9.1 Quantum Reinforcement Learning (QRL) Framework

A quantum reinforcement learning (QRL) framework is an extension of classical reinforcement learning techniques that leverages quantum resources and principles to solve reinforcement learning problems. A QRL framework combines the concepts of quantum computation and quantum mechanics with reinforcement learning algorithms to address challenges and explore potential advantages in certain domains. Here is a general overview of a QRL framework:

A. **Quantum state representation:** In QRL, the state of the environment is represented using quantum states. This can involve encoding the state information into a quantum register or using quantum circuits to represent the state. Quantum state representations offer the potential for enhanced information processing and richer representations compared to classical representations.
B. **Quantum actions and quantum dynamics:** Instead of classical actions, QRL introduces quantum actions, which are operations or transformations performed on the quantum state to affect the environment. These actions can

be represented using quantum gates or quantum circuits. Quantum dynamics describe how the quantum state evolves based on the chosen quantum actions and the underlying physics of the environment.

C. **Quantum rewards and quantum observables:** QRL incorporates quantum rewards, which are measurements or observables associated with the quantum state. These rewards capture the quantum aspects of the environment and can be used to guide the learning process. In QRL, the Q-values, denoted as $Q(s, a)$, represent the expected cumulative rewards for taking action a in state s. These Q-values can be represented using quantum states, such as qubits, where the amplitudes encode the Q-values for each action. The quantum state evolves through quantum operations, and the measurement of qubits provides the Q-values for decision-making.Quantum observables can provide additional information about the state or the environment and can influence the agent's decision-making.
D. **Quantum policy and quantum value functions:** Similar to classical reinforcement learning, QRL involves the use of quantum policies and quantum value functions. A quantum policy defines the agent's behavior and determines which quantum actions to choose based on the current quantum state. Quantum value functions estimate the value or utility of taking a particular quantum action in a given quantum state.
E. **Quantum Q-learning and algorithms:** QRL algorithms aim to find optimal quantum policies and value functions through iterative updates based on the principles of reinforcement learning. These algorithms adapt the quantum actions and policies based on feedback from the environment, including the quantum rewards and observations. Quantum Q-learning is a commonly used algorithm in QRL, but other variations and quantum counterparts of classical RL algorithms may also be employed.
F. **Exploration and exploitation in quantum space:** Exploration-exploitation strategies in QRL involve searching and exploiting the quantum action space to maximize rewards. This can involve quantum exploration techniques, such as quantum randomization or quantum superposition, to probe different quantum actions and learn the optimal policy in the quantum domain.

3.3.9.2 Algorithm of Quantum Reinforcement Learning (QRL)

The algorithm extends the classical Q-learning algorithm to incorporate quantum principles. The algorithm can be summarized as follows:

Input: Environment dynamics E, agent's policy π, discount factor γ, learning rate α, number of episodes N.

Output: Q-values $Q(s, a)$ for each state-action pair.

Algorithm:
Step 1: Initialize Q-values:

- $Q(s, a) = 0$ for all state-action pairs (s, a).

Step 2: For episode = 1 to N:

- Reset the environment to initial state $s0$.
- Repeat until termination:
  - Select an action at using the agent's policy π based on the current state st.
  - Execute action at and observe the next state s_t+1 and the reward rt.
  - Update Q-value:

$$Q(st, at) = (1-\alpha)^{*} Q(st, at) + \alpha^{*}\left[rt + \gamma^{*} \max\_aQ(st+1, a)\right].$$

Step 3: Return the Q-values $Q(s, a)$ for each state-action pair.

Note: The above algorithm is a general Q-learning algorithm, where the Q-values are updated iteratively based on the observed rewards and the maximum Q-value of the next state. The specific implementation details may vary based on the problem and the chosen quantum approach.

3.4 CONCLUSION

Quantum machine learning algorithms hold promise in tackling complex problems by leveraging quantum resources. These algorithms aim to harness the unique properties of quantum systems to enhance data processing, optimization, and pattern-recognition tasks. From quantum support vector machines to quantum neural networks and variational quantum classifiers, researchers are exploring various approaches. While still in its early stages, quantum machine learning has the potential to revolutionize fields like drug discovery, optimization, and quantum information processing. Continued advancements in hardware and algorithm development will further propel the capabilities of quantum machine learning, paving the way for exciting applications and discoveries in the future. In this chapter, we provided a general overview of quantum machine learning, discussing the motivations, benefits, challenges, and key concepts. By exploring the algorithms and concepts in this field, researchers and practitioners can gain insights into the potential of quantum machine learning and contribute to the advancement of this exciting field.

REFERENCES

1. Huang, H.Y., et al. Power of data in quantum machine learning. *Nature Communications* 12, 2631 (2021). https://doi.org/10.1038/s41467-021-22539-9
2. Vedran, Dunjko, Jacob M. Taylor, and Hans J. Briegel. Quantum-enhanced machine learning. *Physical Review Letters* 117 (13), 130501 (2016). https://doi.org/10.1103/PhysRevLett.117.130501
3. Havenstein, Christopher, Damarcus Thomas, and Swami Chandrasekaran. Comparisons of performance between quantum and classical machine learning. *SMU Data Science Review* 1 (4), 11 (2018).

4. Djordjevic, Ivan B. *Chapter 3—Quantum information processing fundamentals, Quantum Communication, Quantum Networks, and Quantum Sensing*, Academic Press (2022), 89–124, ISBN 9780128229422, https://doi.org/10.1016/B978-0-12-822942-2.00008-X
5. Szczepanik, Dariusz W., Emil Zak, and Janusz Mrozek. From quantum superposition to orbital communication. *Computational and Theoretical Chemistry* 1115, 80–87 (2017). ISSN 2210-271X, https://doi.org/10.1016/j.comptc.2017.05.041
6. Havlíček, V., Córcoles, A.D., Temme, K., et al. Supervised learning with quantum-enhanced feature spaces. *Nature* 567, 209–212 (2019). https://doi.org/10.1038/s41586-019-0980-2
7. https://qiskit.org/documentation/stable/0.24/tutorials/machine_learning/01_qsvm_classification.html
8. Hossain, M.M., et al. Analyzing the effect of feature mapping techniques along with the circuit depth in quantum supervised learning by utilizing quantum support vector machine, *2021 24th International Conference on Computer and Information Technology (ICCIT)*, Dhaka, Bangladesh, 2021, 1–5, https://doi.org/10.1109/ICCIT54785.2021.9689853
9. Ezhov, Alexandr A., and Dan Ventura. Quantum neural networks. *Future Directions for Intelligent Systems and Information Sciences: The Future of Speech and Image Technologies, Brain Computers, WWW, and Bioinformatics 1*213–235 (2000).
10. Verdon, Guillaume, Jason Pye, and Michael Broughton. A universal training algorithm for quantum deep learning. arXiv preprint arXiv:1806.09729 (2018).
11. Gonçalves, Carlos Pedro. Quantum neural machine learning-backpropagation and dynamics. arXiv preprint arXiv:1609.06935 (2016).
12. Tilly, Jules, et al. The variational quantum eigensolver: A review of methods and best practices. *Physics Reports* 986, 1–128 (2022), ISSN 0370-1573, https://doi.org/10.1016/j.physrep.2022.08.003
13. Song, Hai-Jing, and D.L. Zhou. Group theory on quantum Boltzmann machine. *Physics Letters A* 399, 127298 (2021), ISSN 0375-9601, https://doi.org/10.1016/j.physleta.2021.127298
14. Lloyd, S., Mohseni, M., and Rebentrost, P. Quantum principal component analysis. *Nature Physics* 10, 631–633 (2014). https://doi.org/10.1038/nphys3029
15. Di Adamo, Stephen, et al. Practical quantum k-means clustering: Performance analysis and applications in energy grid classification. *IEEE Transactions on Quantum Engineering* 3, 1–16 (2022).
16. Dallaire-Demers, Pierre-Luc, and Nathan Killoran. Quantum generative adversarial networks. *Physical Review A* 98 (1), 012324 (2018).
17. Mari, Andrea, et al. Transfer learning in hybrid classical-quantum neural networks. *Quantum* 4, 340 (2020).
18. Dong, Daoyi, et al. Quantum reinforcement learning. *IEEE Transactions on Systems, Man, and Cybernetics, Part B (Cybernetics)* 38 (5), 1207–1220 (2008).

4 Machine Learning with Unsupervised Quantum Models

Prianka Ramachandran Radhabai
New Horizon College of Engineering, India

Sathya Karunanidhi
Vellore Institute of Technology, India

Shreyanth Srikanth
Birla Institute of Technology and Science, India

4.1 INTRODUCTION

Machine Learning (ML) is the process of training machines to analyze and learn from provided data. It can be categorized into three main types: supervised, unsupervised, and reinforcement learning.

Unsupervised learning, a method that allows the discovery of underlying patterns in data without the need for additional information or labeled targets, uncovers hidden patterns within the given dataset but lacks associated labels or classes. Unsupervised techniques prove particularly useful for exploring data and comprehending complex behaviors that challenge human identification within large datasets. They find applications in various fields, such as text categorization (e.g., news articles), anomaly detection, satellite and spatial image processing, medical image analysis, customer segmentation, and recommendation engines.

- Unsupervised learning primarily serves three main tasks: clustering, association, and dimensionality reduction. Clustering is a technique that automatically groups similar samples together based on their inherent characteristics.
- Association is a technique used to discover relationships between different features within a dataset.
- Dimensionality reduction is a technique employed to decrease the number of features in a dataset, especially when dealing with high-dimensional data.

DOI: 10.1201/9781003429654-6

4.2 K-MEANS AND K-MEDIANS CLUSTERING

K-Means clustering is a widely used centroid-based method that divides data points into k predefined clusters. Each data point is assigned to the cluster represented by the nearest centroid. The centroids are updated iteratively by computing the mean of data points in each cluster. The process continues until convergence, producing k clusters based on data point proximity to centroids. It is commonly used for data segmentation and pattern recognition in various fields (Figure 4.1).

K-Means clustering assigns data points to the closest centroid, represented as larger circle markers in the image. Each cluster's centroid is calculated as the mean of its data points. The algorithm iteratively minimizes distances, grouping data into k clusters.

$$C = \frac{1}{Nc} \sum_{xj=1}^{Nc} xj \tag{4.1}$$

where Nc is the number of vectors in the subset. In machine learning, the evaluation of a cost function helps estimate the error in a model's predictions.

Improving a model's performance involves minimizing this function to achieve optimal results. In the context of clustering, the cost function measures the sum of distortions within the clusters. In machine learning, the evaluation of a cost function helps estimate the error in a model's predictions. For clustering, the cost function measures the sum of distortions within the clusters. Specifically, in K-Means clustering, the distortion refers to the squared Euclidean distance between data points and their closest cluster centers. Minimizing this function is crucial to achieving optimal clusters.

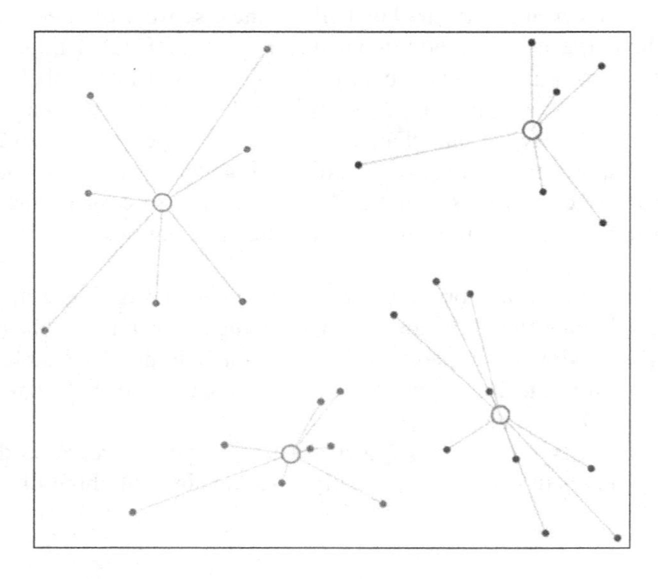

FIGURE 4.1 K-means clustering.

To minimize the cost function J in the K-Means algorithm, two main steps are taken iteratively:

1. Updating the data point assignments to the nearest centroids while keeping the centroids fixed.
2. Updating the centroids based on the newly assigned data points while keeping the assignments fixed.

The K-Median variant of the algorithm uses the Manhattan distance (L1 distance) instead of the Euclidean distance. The Manhattan distance measures the sum of absolute differences between the coordinates of the centroid and a data point. Both K-Means and K-Median algorithms aim to minimize the cost function by iteratively updating data point assignments and centroid positions, with K-Median using the Manhattan distance for distance calculations.

4.2.1 THE K-MEANS ALGORITHM

The K-Means algorithm, also known as Lloyd's algorithm, can be summarized in the following steps:

Step 1: Randomly initialize centroids for each of the k clusters.
Step 2: Assign each data point to the closest centroid to form the initial k clusters.
Step 3: Recompute the centroid by calculating the average of all data points in each of the k clusters. After recomputing the centroids, data points are reassigned to the closest centroid.
Step 4: Repeat Step 3 (recompute centroids and reassign data points) until the data points stop changing clusters.

K-Means implementation in Scikit-Learn has the following key hyperparameters:

1. *n_clusters*: The user provides the number of desired clusters.
2. *init*: Centroids are initialized either randomly or using the K-Means++ technique for better results.
3. *n_init*: The algorithm runs multiple times with different centroid seeds to avoid imbalanced clusters for sparse data, and the best result is selected based on inertia (a clustering performance metric).
4. *max_iter*: The maximum number of iterations allowed for centroid recomputation, useful for controlling processing time.
5. *algorithm*: A choice between Lloyd's or Elkan's algorithm. Lloyd's algorithm is commonly used for K-Means, establishing centroids and partitioning data points into clusters.

Now, let's look at the key attributes available for training:

6. *Cluster_centers_*: An array with the coordinates of the centroids (cluster centers).

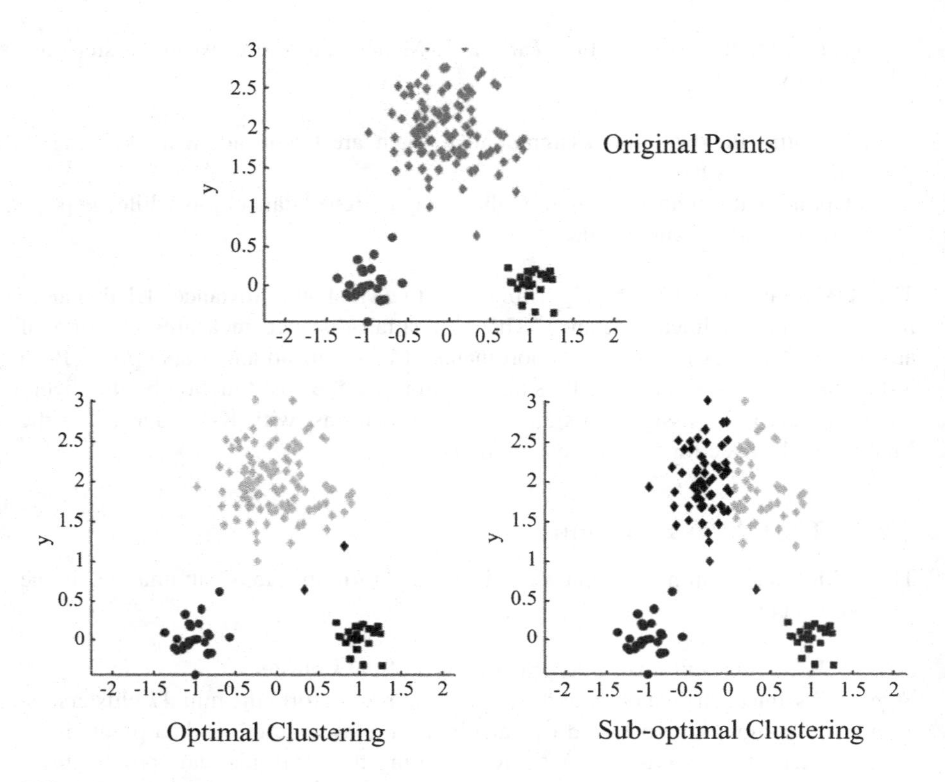

FIGURE 4.2 Different types of K-means clustering.

7. *Labels_*: An array with labels assigned to each data point, representing the cluster they belong to.
8. *Inertia_*: A metric that measures how well the clusters are formed, calculated as the sum of squared distances of each sample to its closest cluster center.

Note: K-Means is an unsupervised learning model and is fitted only on the training data. The fit, fit_transform, and fit_predict methods of K-Means take only one argument, which is the dataset to be observed and clustered (Figure 4.2).

Problem: Given a set X of n points in a d-dimensional space and an integer k, the objective is to group the points into k clusters $C = \{C1, C2, \ldots, Ck\}$ such that:

$$\text{Cost}(C) = \sum_{i=1}^{k} \sum_{x \in C_i} dist(x, c) \tag{4.2}$$

is minimized, where ci is the centroid of the points in cluster C_i.

The most common definition is with Euclidean distance, minimizing the Sum of Squared Error (SSE) function.

Problem: Given a set *X* of *n* points in a d-dimensional space and an integer *k*, the task is to group the points into k clusters $C = \{C1, C2, \ldots, Ck\}$ such that:

$$\text{Cost}(C) = \sum_{i=1}^{k} \sum_{x \in C_i} (x - c_i)^2 \tag{4.3}$$

is minimized, where *ci* is the mean of the points in cluster C_i.

Advantages of K-Means

9. *Simple and easy to implement*: The K-Means algorithm is straightforward to understand and use, making it a popular choice for clustering tasks.
10. *Fast and efficient*: K-Means is computationally efficient, making it suitable for large datasets with high dimensionality.
11. *Scalability*: K-Means can handle large datasets with a large number of data points and can be easily scaled to handle even larger datasets.
12. *Flexibility*: K-Means is versatile and can be adapted to different applications, allowing the use of various distance metrics and initialization methods.

Disadvantages of K-Means

13. *Sensitivity to initial centroids*: K-Means can converge to a suboptimal solution because it is sensitive to the initial selection of centroids.
14. *Requires specifying the number of clusters*: K-Means requires knowing the number of clusters (k) beforehand, which can be difficult to determine in some applications.
15. *Sensitive to outliers*: K-Means is affected by outliers, and their presence can lead to skewed or inaccurate cluster assignments.

Applications of K-Means Clustering

K-Means clustering is used in a variety of examples or business cases in real life, such as:

16. *Academic performance:* K-Means can be used for student performance analysis and grouping students based on their academic achievements or study patterns.
17. *Diagnostic systems:* In medical applications, K-Means can be employed for disease diagnosis, clustering patients based on their symptoms and medical data.
18. *Search engines:* K-Means can help in document clustering for search engines, grouping similar documents to enhance search results.
19. *Wireless sensor networks*: K-Means is used to organize sensor data in wireless sensor networks for efficient data processing and analysis.

4.3 HIERARCHICAL AND DENSITY-BASED CLUSTERING

4.3.1 Hierarchical Clustering

Hierarchical clustering is a flexible method that forms a hierarchy of clusters (dendrogram) without requiring a fixed number of clusters beforehand. Divisive clustering is performed when the number of clusters increases, where all data instances start in one cluster and split in each iteration, resulting in a hierarchy of clusters. The algorithm includes single linkage, complete linkage, average linkage, and Ward's method, which form a set of nested clusters organized in a hierarchical tree.

The hierarchical clustering algorithm is of two types. Agglomerative hierarchical clustering (AGNES) is an algorithm that groups data points based on their pairwise distance measures. It starts by treating each data point as a separate cluster and then successively merges the closest pairs of clusters until a single cluster remains. The distance between clusters can be computed using various methods, including single-nearest distance (single linkage), complete-farthest distance (complete linkage), average-average distance (average linkage), centroid distance, and Ward's method (minimizing sum of squared Euclidean distances). Conversely, the divisive method (DIANA) starts with all data points in one cluster and iteratively divides them into smaller clusters. Both methods are employed in hierarchical clustering to construct dendrograms and identify the optimal number of clusters.

4.3.1.1 Agglomerative Hierarchical Clustering

Agglomerative hierarchical clustering (AGNES) proceeds with the data grouping until only one cluster is formed. To determine the appropriate number of clusters, the dendrogram graph is analyzed based on specific criteria. It is worth noting that AGNES is an agglomerative approach, where it begins with individual data points and combines them into larger clusters.

Algorithmic steps for agglomerative hierarchical clustering:

Let $X = \{x1, x2, x3, \ldots, xn\}$ be the set of data points.

4.3.1.2 Divisive Hierarchical Clustering

Divisive hierarchical clustering (DIANA) is the reverse of AGNES, where the former starts with one large cluster and divides it into smaller clusters in each iteration, while the latter begins with individual data points and progressively merges them into larger clusters.

Divisive hierarchical clustering is a hierarchical approach that begins with all data points in a single cluster and then iteratively divides them into smaller clusters. The steps involved in this process are as follows:

20. Start with a single cluster labeled as $m = 0$ and having level $L(0) = 0$.
21. Identify the two clusters with the greatest distance between their data points in the current clustering, represented as pair (r) and (s), based on $d[(r), (s)] = \min d[(i), (j)]$ where the minimum is taken over all pairs of clusters in the current clustering.

22. Increment the sequence number: $m = m + 1$. Divide clusters (r) and (s) to create two separate clusters, forming the next clustering labeled as m. Set the level of this clustering to $L(m) = d[(r), (s)]$.
23. Update the distance matrix, D, by removing the rows and columns corresponding to clusters (r) and (s) and adding a row and column for the newly formed clusters. The distance between the new clusters, denoted as (r, s), and an old cluster (k) is defined as follows: $d[(k), (r, s)] = \min(d[(k), (r)], d[(k), (s)])$.
24. If all the data points are now in distinct clusters, stop the process. Otherwise, repeat from Step 2.

Advantages of Hierarchical Clustering

1. No need for prior information about the number of clusters.
2. Simple to implement and can yield optimal results in certain cases.

Divisive hierarchical clustering starts with a single, all-inclusive cluster and proceeds to split clusters one by one until each cluster contains a single data point (or a specified number of clusters, k). It uses a similarity or distance matrix to determine cluster similarities. The algorithm merges or splits clusters iteratively, creating a hierarchical tree structure known as a dendrogram. This tree-like diagram records the sequences of merges or splits, providing a visualization of the clustering process and the nested clusters (Figure 4.3).

We do not have to assume any particular number of clusters:

- Any desired number of clusters can be obtained by "cutting" the dendrogram (see Figure 4.4) at the proper level.
- Clusters may correspond to meaningful taxonomies.
- There are many examples in biological sciences (e.g., animal kingdom, phylogeny reconstruction, etc.)

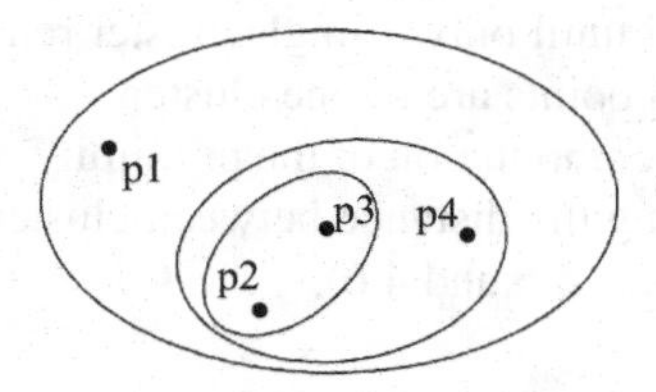

FIGURE 4.3 Traditional hierarchical clustering.

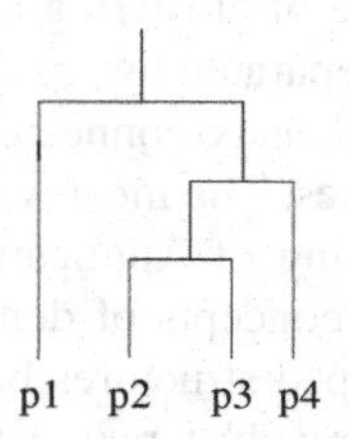

FIGURE 4.4 Traditional dendrogram.

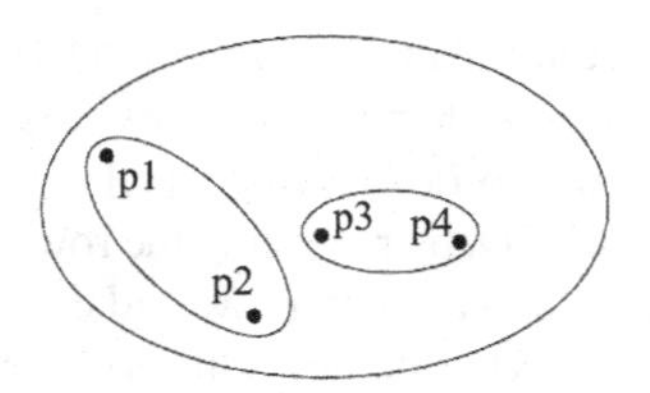

FIGURE 4.5 Non-traditional hierarchical clustering.

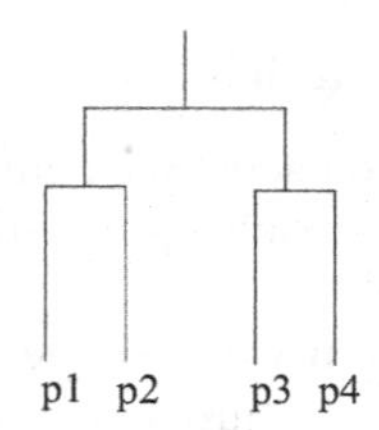

FIGURE 4.6 Non-traditional dendrogram.

The most popular hierarchical clustering technique:

25. Compute the proximity matrix: Calculate the distances between all data points.
26. Treat each data point as a separate cluster: Start with each data point as its own cluster.
27. Merge the two closest clusters: Repeatedly merge the two clusters with the smallest distance.
28. Update the proximity matrix: Recalculate distances between the new merged cluster and the remaining clusters.
29. Repeat Steps 3 and 4 until only a single cluster remains: Continue merging clusters until all data points are in one cluster.
30. Key operation is the computation of the proximity of two clusters. Different approaches to defining the distance between clusters distinguish the different algorithms (Figures 4.5 and 4.6).

4.3.2 Density-Based Clustering

Density-based clustering is a type of clustering algorithm that identifies clusters as dense regions in the data space, separated by regions of lower object density. A cluster is defined as a maximal set of density-connected points, allowing the algorithm to discover clusters of arbitrary shapes. The most widely used density-based algorithm is Density-Based Spatial Clustering of Applications with Noise (DBSCAN). The DBSCAN algorithm utilizes the concepts of density reachability and density connectivity to find non-linearly shaped structures based on data density. This method has been instrumental in identifying clusters in datasets with varying densities and is particularly effective at handling noisy data points (Figure 4.7).

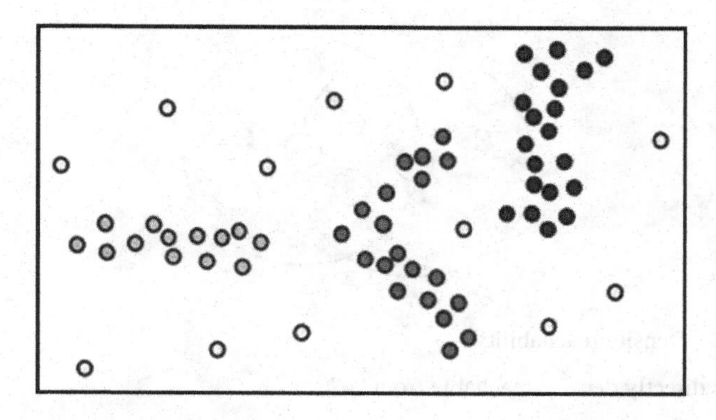

FIGURE 4.7 Density-based spatial clustering.

4.3.2.1 Density Definition

ε-Neighborhood – Objects within a radius of ε from an object.

$$N_{\varepsilon}(p):\{q \mid d(p,q) \leq \varepsilon\} \tag{4.4}$$

"High density" – ε-Neighborhood of an object contains at least MinPts of objects (Figure 4.8).

4.3.2.2 Density Reachability

A point "p" is considered to be density reachable from a point "q" if the following conditions are met:

31. Point "p" is within a specified distance ε (epsilon) from point "q".
32. Point "q" has a sufficient number of points in its ε-neighborhood (i.e., within a distance ε from "q").

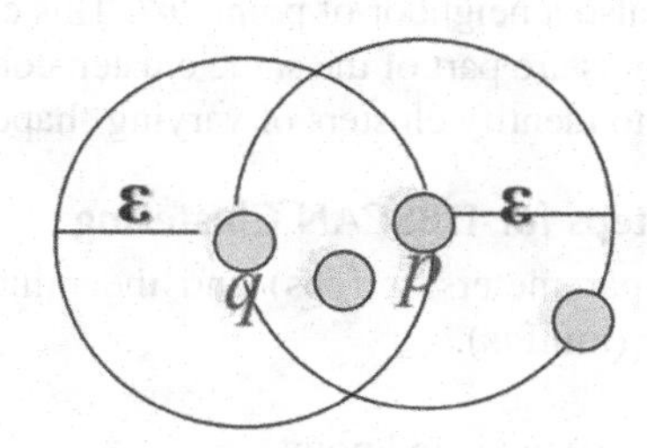

FIGURE 4.8 Density definition.

ε-Neighborhood of p
ε-neighborhood of q
Density of p is "high" (MinPts = 4)
Density of q is "low" (MinPts = 4)

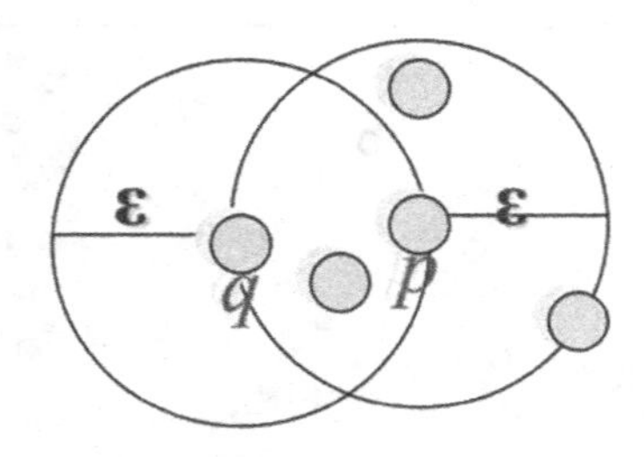

FIGURE 4.9 Density reachability.

where, "q" is directly density-reachable from "p".
MinPts = 4
p is not directly density-reachable from q
Density-reachability is asymmetric

In other words, for point "p" to be density reachable from "q", "q" should have enough neighboring points within ε distance, indicating a higher density region. This notion of density reachability allows the algorithm to identify connected and dense clusters in the data space, even if the clusters have irregular shapes or varying densities (Figure 4.9).

4.3.2.3 Density Connectivity

Two points "p" and "q" are considered to be density connected if the following conditions are satisfied:

33. There exists a point "r" that has a sufficient number of points in its ε-neighborhood.
34. Both points "p" and "q" are within the ε distance from "r".

This process is often referred to as chaining, where density-connected points form a chain or path of neighboring points. So, if point "q" is a neighbor of point "r", "r" is a neighbor of point "s", "s" is a neighbor of point "t", and "t" is a neighbor of point "p", it implies that "q" is also a neighbor of point "p". This chain-like process ensures that density-connected points are part of the same cluster, contributing to the ability of density-based algorithms to identify clusters of varying shapes and sizes in the data.

4.3.2.4 Algorithmic Steps for DBSCAN Clustering

DBSCAN requires two parameters: ε (eps) and the minimum number of points required to form a cluster (minPts).

35. Begin with an arbitrary unvisited point.
36. Form the neighborhood of this point using ε (including all points within ε distance).
37. If the neighborhood contains enough points, start the clustering process, mark the point as visited, otherwise label it as noise (which may join a cluster later).

38. If a point is part of the cluster, its ε neighborhood is also part of the cluster, and the process repeats from Step 2 for all the ε neighborhood points. This continues until all points in the cluster are determined.
39. Retrieve and process a new unvisited point, leading to the discovery of another cluster or noise.
40. Continue the process until all points are marked as visited.

Advantages of DBSCAN:

41. No need to specify the number of clusters beforehand.
42. Can detect and label noisy data as outliers.
43. Can find clusters of various shapes and sizes.

Disadvantages of DBSCAN:

44. Not suitable for datasets with varying density clusters.
45. Struggles with "neck-like" datasets.
46. Performance may degrade with high-dimensional data.

4.4 FUZZY C-MEANS CLUSTERING

Fuzzy C-Means clustering is an extension of K-Means clustering. Unlike K-Means, where each data point belongs to only one cluster, Fuzzy C-Means allows data points to be assigned to multiple clusters with membership degrees (probabilities) between 0 and 1. In Fuzzy C-Means clustering, a data point can belong to every cluster with a certain weight, representing its membership strength. The algorithm assigns membership values based on the distance between data points and cluster centers. After each iteration, memberships and cluster centers are updated, and this process continues until convergence, providing more flexible clustering and handling uncertainty in the data.

$$\mu_{ij} = 1 \Big/ \sum_{k=1}^{c} \left(d_{ij} / d_{ik} \right)^{(2/m-1)} \tag{4.5}$$

$$v_j = \left(\sum_{i=1}^{n} \left(\mu_{ij} \right)^m x_i \right) \Big/ \left(\sum_{i=1}^{n} \left(\mu_{ij} \right)^m \right), \quad \forall j = 1,2,\ldots c \tag{4.6}$$

where, "n" is the number of data points. "v_j" represents the jth cluster center. "m" is the fuzziness index m € [1, ∞]. "c" represents the number of cluster center. "μ_{ij}" represents the membership of ith data to jth cluster center. "d_{ij}" represents the Euclidean distance between ith data and jth cluster center.

The main objective of the Fuzzy C-Means algorithm is to minimize:

$$J(U,V) = \sum_{i=1}^{n} \sum_{j=1}^{c} \left(\mu_{ij} \right)^m \left\| x_i - v_j \right\|^2 \tag{4.7}$$

where, "$||x_i - v_j||$" is the Euclidean distance between ith data and jth cluster center.

4.4.1 Algorithmic Steps For Fuzzy C-Means Clustering

Let $X = \{x1, x2, x3 \ldots, xn\}$ be the set of data points and $V = \{v1, v2, v3 \ldots, vc\}$ be the set of centers.

1. Start by randomly selecting "c" cluster centers.
2. Calculate the fuzzy membership "μ_{ij}" using the following formula:

$$\mu_{ij} = 1 \Big/ \sum_{k=1}^{c} \left(d_{ij} / d_{ik} \right)^{(2/m-1)} \tag{4.8}$$

3. Compute the fuzzy centers "v_j" using:

$$v_j = \left(\sum_{i=1}^{n} \left(\mu_{ij} \right)^m x_i \right) \Big/ \left(\sum_{i=1}^{n} \left(\mu_{ij} \right)^m \right), \quad \forall j = 1, 2, \ldots\ldots c \tag{4.9}$$

4. Repeat Steps 2 and 3 until the minimum "j" value is achieved or $||U(k+1) - U(k)|| < \beta$.

where,

"k" is the iteration step.
"β" is the termination criterion between [0, 1].
"$U = (\mu_{ij})n*c$" is the fuzzy membership matrix.
"j" is the objective function (Figure 4.10).

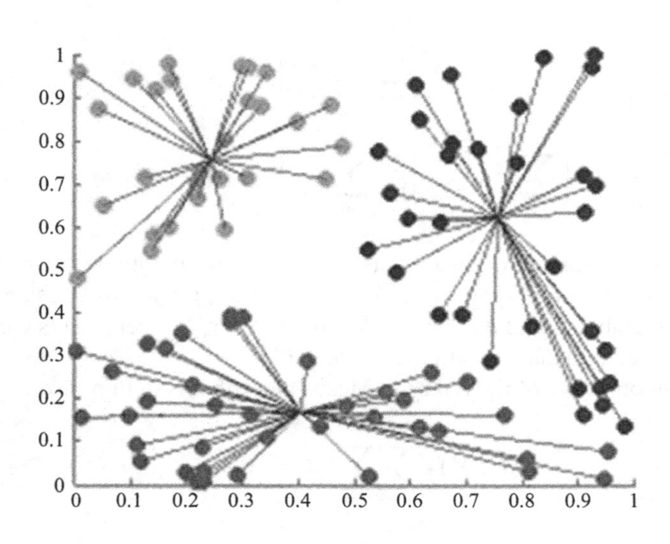

FIGURE 4.10 Result of fuzzy C-means clustering.

Advantages of Fuzzy C-Means Clustering:

47. Provides superior results for overlapped datasets and generally outperforms the K-Means algorithm in such cases.
48. Allows data points to be assigned memberships to multiple cluster centers, accommodating the possibility of data points belonging to more than one cluster.

Disadvantages of Fuzzy C-Means Clustering:

49. Requires the specification of the number of clusters beforehand.
50. Achieving better results with lower values of β may lead to a higher number of iterations, affecting computational efficiency.
51. Euclidean distance measures may not equally weigh underlying factors, potentially affecting the accuracy of the clustering results.

4.4.2 Clustering Algorithm Applications

4.4.2.1 Clustering Algorithms in Identifying Cancerous Data

Clustering algorithms can be effectively utilized to identify cancerous datasets. The process involves taking known samples of both cancerous and non-cancerous data and labeling them accordingly. These labeled samples are then mixed randomly to create a combined dataset, and various clustering algorithms are applied to this mixed dataset during the learning phase.

During the learning phase, the clustering results are compared with the known labels of the original samples to determine how many correct results are obtained. This allows us to calculate the percentage of correct results achieved by each algorithm. Based on this analysis, we can identify the most suitable clustering algorithm for our data samples.

Once the best-performing algorithm is determined, it can be applied to new and arbitrary samples of data to expect a similar percentage of correct results as obtained during the learning phase. For cancerous datasets, it has been observed through experimentation that unsupervised non-linear clustering algorithms yield the best results. This finding suggests the non-linear nature of the cancerous data, emphasizing the significance of unsupervised non-linear clustering techniques in cancer data analysis.

4.4.2.2 Clustering Algorithms in Search Engines

Clustering algorithms form the foundation of search engines. They group similar objects together and separate dissimilar ones, making search results more relevant and useful. The quality of the clustering algorithm used in a search engine greatly affects the presentation of search results. A good clustering algorithm can improve the chances of getting desired results on the front page. To achieve better search results, it's crucial to define the criteria for determining similar objects. Brainstorming and careful consideration are needed for this process. The definition of similar objects plays a vital role in search engine performance. The better the definition, the more accurate and relevant the search results will be.

4.4.2.3 Clustering Algorithms in Academics

The ability to monitor the progress of students' academic performance has been the critical issue for the academic community of higher learning. Clustering algorithms can be used to monitor the students' academic performance. Based on the students' score they are grouped into different-different clusters (using K-Means, Fuzzy C-Means, etc.), where each cluster denotes the different level of performance. By knowing the number of students' in each cluster we can know the average performance of a class as a whole.

4.4.2.4 Clustering Algorithms in Wireless Sensor Network-Based Application

Clustering algorithms are effective in Wireless Sensor Network (WSN) applications, such as landmine detection. In landmine detection, clustering algorithms help identify cluster heads (or cluster centers) responsible for collecting data within their respective clusters. cluster heads in WSNs efficiently organize data collection, reduce communication overhead, and optimize energy consumption. By using clustering algorithms, WSNs can improve accuracy and reliability in identifying potential landmine locations, making them more effective in critical tasks like landmine detection.

4.4.3 Example Problems

Activity:

- Explain the Fuzzy C-Means Clustering (FCMC) technique
- Try to cluster data points into fuzzy clusters using the FCMC technique

Given a set of data, Clustering techniques are used to partition a set of data into multiple groups, ensuring strong association within each group and weak association between data points in different groups. Classical crisp clustering techniques create crisp partitions, where each data point belongs to only one cluster. An illustration of this is shown in Figure 4.11 where six data points are clustered into two distinct clusters. In this example, the data has two dimensions, meaning it exists in a two-dimensional feature space (Figure 4.11).

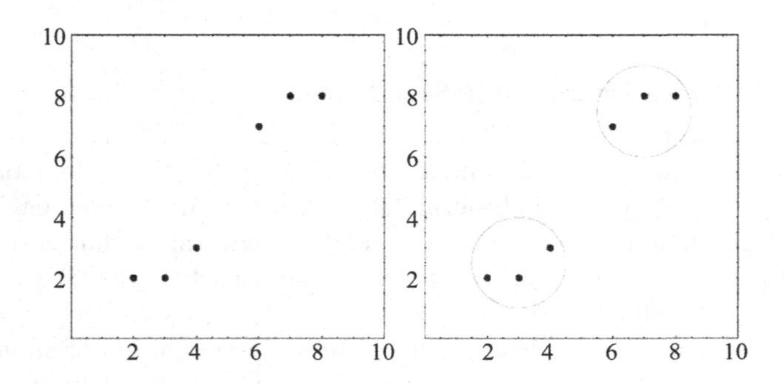

FIGURE 4.11 A simple example for crisp clustering.

Fuzzy clustering, in contrast to crisp clustering, allows data points to belong to multiple groups, resulting in a fuzzy partition. Each cluster is associated with a membership function that quantifies the degree of membership of individual data points within the cluster. Among various fuzzy clustering methods, Fuzzy C-Means clustering (FCMC) is the most widely used and researched in both academic and industrial applications [1]. Its success and popularity have made it a predominant technique in the literature.

Activity:

- Observe that the number of clusters should be given in advance.
- Observe the iterativity in the FCMC algorithm.

Fuzzy C-Means clustering performs clustering by iteratively searching for a set of fuzzy clusters and their associated cluster centers, which best represent the data's structure. The user needs to specify the number of clusters (c) present in the data set to be clustered. FCMC then partitions the data, $X = \{x1, x2, \ldots, xn\}$, into c fuzzy clusters by minimizing the within-group sum of squared error objective function as follows:

$$J_m(U,V) = \sum_{k=1}^{n}\sum_{i=1}^{c}(U_{ik})^m \|x_k - v_i\|^2, \quad 1 \le m \le \infty \tag{4.10}$$

where $J_m(U, V)$ represents the sum of squared error for the set of fuzzy clusters represented by the membership matrix U and the associated cluster centers V. The notation $||.||$ denotes a specific inner product induced norm. In this context, $||x_k - v_i||^2$ represents the distance between a data point x_k and the cluster center v_i. The squared error serves as a performance index, measuring the weighted sum of distances between cluster centers and elements in the corresponding fuzzy clusters. The parameter "m" governs the influence of membership grades in the performance index. As "m" increases, the partition becomes fuzzier. The FCMC algorithm has been proven to converge for any value of "m" within the range of $(1, \infty)$.

In each iteration of the FCMC algorithm, matrix U is computed and the associated cluster cents are computed as Eq. (4.6). This is followed by computing the square error in Eq. (4.5). The algorithm stops when either the error is below a certain tolerance value or its improvement over the previous iteration is below a certain threshold. The clustering process is displayed in Figure 4.12 (initial state) and Figure 4.13 (final state) using three clusters.

Activity:

- See the different types of clustering algorithms, note their advantages. Over the years, numerous extensions and variations of Fuzzy C-Means clustering (FCMC) have been proposed. By using different distance functions in Eq. (4.5) or making slight modifications to the objective function, these clustering algorithms have become capable of detecting different types of clusters.

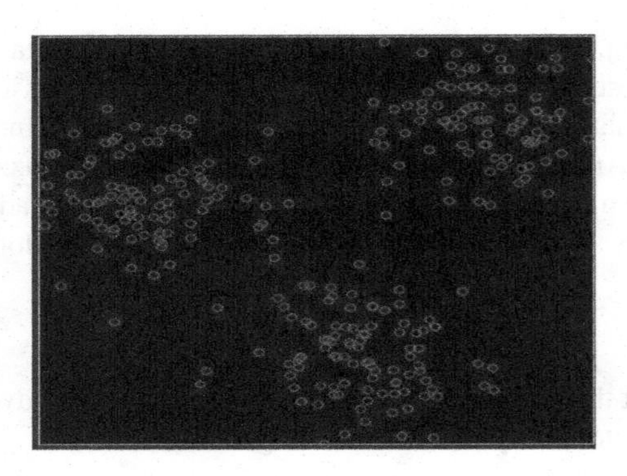

FIGURE 4.12 Clustering process (initial state).

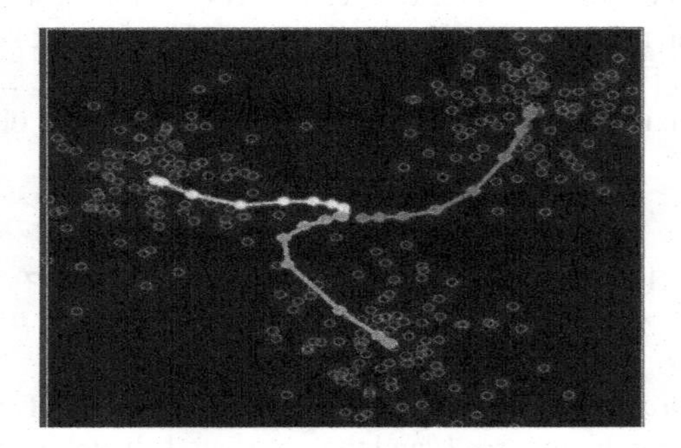

FIGURE 4.13 Clustering process (final state).

When FCMC uses the Euclidean distance, it becomes effective in detecting approximately similar-sized spherical clusters. Gustafson and Kessel introduced the GK clustering algorithm [2], which uses a transformed Mahalanobis distance. This modification allows GK clustering to detect cylinder-shaped normal clusters of approximately the same size. The effectiveness of the GK clustering technique was analyzed and compared to other clustering methods in [3]. The distance formula used in the GK clustering algorithm is given by Eq. (4.8), where C_i represents the covariance matrix for the i-th cluster, $C_i - 1$ denotes the inverse of the covariance matrix, d is the number of dimensions, and $\rho_i = 1$ is a constant.

4.5 PRINCIPAL COMPONENT ANALYSIS FOR QUANTUM COMPUTING

Principal Component Analysis (PCA) is a powerful dimensionality reduction approach with applications in a wide range of industries, including quantum computing. Dealing with high-dimensional quantum data is a common difficulty in quantum computing, which is a fast-expanding field. However, PCA provides an effective method for extracting critical features and reducing the complexity of quantum datasets, resulting in better quantum algorithms and analysis.

In the context of Quantum Computing, PCA entails determining the primary components of quantum states or quantum data. These primary components describe the directions in the quantum feature space that demonstrate the most substantial variability in the data. We can minimize the dimensionality of the data while keeping its key information by projecting the quantum data onto these principal components.

The mathematical roots of PCA in Quantum Computing are based on linear algebra principles such as eigenvalue decomposition and singular value decomposition (SVD). Density matrices can be used to represent quantum states or quantum data, and their covariance matrix can be constructed to extract the primary components.

The implementation of quantum PCA methods varies, with some using variational quantum circuits to discover approximate answers and others using quantum versions of SVD techniques. These methods bring up new avenues for quantum data processing, such as quantum state compression, noise reduction, and quantum machine learning.

One of the key benefits of PCA in quantum computing is its ability to aid in quantum error mitigation. It can help in discovering and mitigating mistakes by lowering the dimensionality of quantum data, hence improving the overall reliability and performance of quantum calculations.

Furthermore, Quantum PCA is used in quantum chemistry and material science, where high-dimensional quantum states are frequently encountered. Researchers can get deeper insights into complicated quantum systems and better quantum algorithms for simulating chemical interactions and material properties by efficiently expressing these states using PCA.

Despite its benefits, Quantum PCA has limits. Classical simulations to compute PCA get difficult when the amount of the quantum dataset grows due to the exponential expansion of computer resources required. Furthermore, quantum noise and restricted qubit coherence make estimating principal components in quantum hardware difficult.

4.5.1 Variational Quantum Principal Component Analysis

Variational Quantum Principal Component Analysis (VPCA) is based on a variational technique that uses quantum circuits to approximate the primary components of quantum states or quantum datasets. The power of quantum parallelism is used to efficiently search the quantum feature space and locate the paths that capture the most important variability in the data.

The VPCA algorithm starts by converting the quantum data into a quantum state, which is subsequently processed by a parameterized quantum circuit. Classical optimization techniques are used to iteratively tune the circuit's parameters in order to minimize a cost function that quantifies the difference between the input quantum data and the reconstructed quantum state using the selected primary components.

One of the primary benefits of VPCA is its ability to be implemented on near-term quantum devices, when full-scale quantum algorithms may not yet be practical due to hardware constraints. Even in the face of noise and flaws in quantum hardware, VPCA is a useful tool for quantum data analysis and feature extraction.

In addition, VPCA has a lot of promise for quantum machine learning applications. It enables efficient and effective grouping, classification, and other quantum data-driven tasks by lowering the dimensionality of quantum data, thus opening up new paths for research in quantum-enhanced machine learning.

However, it is critical to recognize VPCA's problems and limitations. Its efficacy as a near-term quantum algorithm is strongly dependent on the capabilities and stability of present quantum hardware. In quantum systems, noise and decoherence can reduce the accuracy of the derived principal components, necessitating effective error mitigation techniques.

4.5.2 Quantum Singular Value Decomposition

Quantum Singular Value Decomposition (QSVD) is a key achievement in the field of Quantum Computing, and it plays an important part in the chapter titled "PCA in Quantum Computing." It is a quantum-inspired variant of the traditional Singular Value Decomposition (SVD) algorithm, which is commonly used in classical computing for Principal Component Analysis (PCA).

Singular Value Decomposition (SVD) is used in conventional PCA to divide a matrix into three component matrices, which allows the extraction of main components and their related singular values. Similarly, QSVD decomposes quantum data in a quantum-inspired manner, allowing the identification of important quantum features and patterns.

The QSVD algorithm works by taking advantage of the quantum features of superposition and entanglement. It can efficiently process quantum states and quantum datasets, even in high-dimensional areas. QSVD surpasses classical SVD in terms of speed when applied to quantum data by utilizing quantum parallelism, which is especially helpful for quantum algorithms that deal with big datasets.

The implementation of QSVD into PCA for Quantum Computing offers tremendous promise in the context of the book chapter. It can be used to decrease the dimensionality of quantum states, making them easier to analyze and manipulate. Furthermore, the singular values acquired using QSVD can aid in approximating quantum states, allowing for quantum state compression and noise reduction.

However, in real applications, it is critical to recognize QSVD's limits. Noise, mistakes, and decoherence in quantum hardware can all have an impact on the QSVD algorithm's accuracy. To achieve accurate results in quantum computing applications, a detailed grasp of error mitigation strategies and quantum error models is required.

4.6 ANOMALY DETECTION USING QUANTUM TECHNOLOGIES

Data that differ from typical data patterns are found via anomaly detection. Its application to traditional data has a wide range of uses in many crucial fields, including fraud detection, medical diagnosis, data cleaning, and surveillance. The emergence of quantum technologies may lead to the development of quantum applications that heavily rely on anomaly detection of quantum data in the form of quantum states. For some issues, including factoring and searching an unstructured database, quantum computing has been successful in developing algorithms that are faster than their classical counterparts. It makes use of methods like amplitude amplification, quantum matrix inversion, and quantum phase estimation. Recently, these technologies have been used in machine learning quantum algorithms.

Anomaly detection is one of the many industries that quantum computing could revolutionize. While anomaly detection tasks can be accomplished using classical computers, quantum computing has potential advantages in terms of computational power and the capacity to process massive volumes of data at once.

Quantum machine learning techniques are one of the potential uses of quantum computing in anomaly identification. These algorithms conduct complex computations more quickly than their classical equivalents by taking advantage of the special characteristics of quantum systems.

Algorithms for quantum machine learning may boost anomaly detection by increasing the precision and speed of data analysis. For instance, by transforming the input data into a higher-dimensional feature space and identifying an ideal hyperplane that divides the two classes, quantum support vector machines (QSVMs) can be used to categorize data points as normal or anomalous. These high-dimensional mapping computations can be sped up and the classification process optimized using quantum methods.

In order to solve optimization issues, quantum computing may also be useful for the detection of anomalies. Numerous anomaly detection methods rely on the optimization of particular objective functions to find anomalies. In comparison to conventional optimization techniques, quantum algorithms, such as quantum annealing or quantum-inspired algorithms, may be able to tackle these optimization problems more quickly and effectively.

It is crucial to keep in mind that quantum computing is still in its infancy and that practical quantum machines with adequate qubit counts and error-correction capabilities are still not readily accessible. The use of quantum computing for anomaly detection is, therefore, still mostly theoretical and experimental. Anomaly detection is only one of the many applications for which researchers are actively investigating quantum algorithms and creating methods to harness the power of quantum computing.

Machine learning anomaly detection methods come in many different forms. Each method can be divided into three categories. Depending on variables like the type of data, each sort of approach will incorporate particular outlier detection and analysis algorithms and approaches. The general premise behind each technique is that anomalies are uncommon and markedly dissimilar from the characteristics of typical data points. The following are typical methods for machine learning anomaly detection:

52. Unsupervised anomaly detection.
53. Supervised anomaly detection.
54. Semi-supervised anomaly detection.

4.6.1 Unsupervised Anomaly Detection

The practice of locating aberrant patterns or outliers in a dataset without the aid of labeled data or prior knowledge of the anomalies is known as unsupervised anomaly detection. When there are no pre-existing samples or labels available for training a supervised anomaly detection model, it is a beneficial strategy.

Unsupervised anomaly detection can be done in a number of ways.

- *Approaches based on statistics*: These approaches presuppose that normal data points adhere to a particular statistical distribution, such as the Gaussian distribution. Then, data points that considerably depart from the predicted distribution are recognized as anomalies. Z-scores, percentiles, and multivariate statistical approaches like Mahalanobis distance are a few examples of statistical procedures.
- *Density-based techniques*: These techniques seek to find anomalies in areas with low data density. Anomalies are data points that are located in sparser areas of the dataset. The DBSCAN (Density-Based Spatial Clustering of Applications with Noise) algorithm is one well-known density-based algorithm.
- *Clustering-based approaches*: Anomalies are data points that do not belong to any cluster or only belong to small clusters. These methods attempt to arrange similar data points into clusters. The k-means algorithm is a popular clustering algorithm used for anomaly identification.
- *Autoencoder's neural network models*: called autoencoders are trained to discover a compressed representation of the input data. By contrasting the autoencoder's reconstruction error, anomalies can be found. Higher reconstruction errors increase the likelihood of abnormal data points.
- *Isolation Forest*: Isolation Forest is an ensemble-based technique that recursively partitions the feature space and isolates data points at random. Data points that need fewer partitions to be isolated are detected as anomalies, highlighting their uniqueness.

It's important to keep in mind that unsupervised anomaly detection methods may have drawbacks, such as the inability to clearly define an anomalous detection threshold or the possibility of false positives. Therefore, to increase the precision and dependability of unsupervised anomaly detection systems, it is frequently necessary to integrate different algorithms or to include domain knowledge.

4.6.2 Supervised Anomaly Detection

The practice of locating aberrant patterns or outliers in a dataset utilizing labeled data or prior knowledge of the anomalies is known as supervised anomaly detection. It entails employing a model that has been trained on a labeled dataset with explicitly identified or labeled anomalies to categorize new, unforeseen data points as anomalous or normal. Usually, supervised anomaly detection involves the following steps:

55. *Data gathering and labeling*: Compile a dataset with both typical and unusual cases. Anomalies can be classified using manual inspection, historical data, or specialist knowledge. The dataset needs to reflect the actual environmental factors that the anomaly detection algorithm will be applied to.

56. *Selection and extraction of features*: Find the data's pertinent characteristics or traits that can be used to differentiate between expected and unexpected patterns. In this stage, raw data are converted into a feature representation that the anomaly detection model may use.
57. *Train a supervised learning model using the labeled dataset, such as a binary classifier*: Based on the labeled instances, the model learns to distinguish between typical and abnormal patterns. Support vector machines (SVMs), random forests, or neural networks are supervised learning techniques that are frequently used for anomaly identification.
58. *Validation and model assessment*: Utilize evaluation metrics, such as accuracy, precision, recall, or F1 score, to rate the trained model's performance. To make sure the model can generalize, cross-validation approaches can be used.
59. *Deployment and inference*: The model can be used to forecast abnormalities in unobserved data after it has been trained and assessed. Based on the expertise gained during training, the model labels each data point as either normal or abnormal.

Using labeled data or prior knowledge of anomalies can help in supervised anomaly detection. However, it makes the supposition that the samples with labels adequately depict the abnormalities seen in the actual data. Furthermore, the sorts of anomalies included in the training data might be a constraint on the model's performance. As new anomalies appear or current ones alter, it is crucial to reevaluate and update the model on a regular basis.

4.6.3 Semi-Supervised Anomaly Detection

A strategy that includes aspects of both supervised and unsupervised anomaly detection is called semi-supervised anomaly detection. To train a model that can spot anomalies in unobserved data, it uses a small amount of labeled anomaly data together with a larger amount of unlabeled data. The steps that are commonly involved in semi-supervised anomaly detection are as follows:

60. *Labeled data collection*: Compile a small number of data points with labels that are recognized as abnormalities. These labeled abnormalities may be discovered through manual inspection, historical documents, or professional knowledge.
61. *Gathering of a larger group of unlabeled data points*: These unlabeled data points lack any labels that specifically state whether they are normal or abnormal. They indicate the typical patterns or behaviors in the dataset.
62. *Feature extraction and selection*: From both labeled and unlabeled data points, extract and choose pertinent features. These traits ought to reflect the qualities that set anomalies apart from regular patterns.
63. *Model training*: Using the labeled anomalies and the unlabeled normal data, build a semi-supervised learning model. This model learns to detect abnormalities in the unlabeled data by generalizing from the labeled anomalies. Self-training, co-training, and generative models like autoencoders are common semi-supervised learning algorithms used for anomaly identification.

64. *Model evaluation*: Use the right evaluation metrics to gauge the trained model's performance. Since the unlabeled data lacks clear labels, evaluation may involve comparing the model's anomaly rankings or scores with expert domain knowledge or employing anomaly detection evaluation approaches like precision at a given recall.

The benefit of semi-supervised anomaly detection is that it can identify anomalies in the unlabeled data while simultaneously using a limited quantity of labeled data to direct the learning process. In situations where gathering labeled anomaly data is costly, time-consuming, or not easily available, this method may be helpful.

It is crucial to remember that the caliber and representativeness of the labeled anomalies and the unlabeled normal data are crucial to the success of semi-supervised anomaly detection. The capacity of the model to generalize to hidden abnormalities may be constrained by incomplete or biased labeled data. To get the best results, care should be used in the selection and balancing of labeled and unlabeled data. Types of anomaly detection algorithms are as follows:

- Classification-based anomaly detection algorithms
- Spectral theory-based anomaly detection algorithms
- Nearest neighbor-based anomaly detection algorithms
- Cluster-based anomaly detection algorithms
- Statistical techniques anomaly detection algorithms
- Information-theoretic techniques anomaly detection algorithms

4.6.3.1 Classification-Based Anomaly Detection Algorithms

Classification-based anomaly detection algorithms aim to classify data points as either normal or anomalous by training a classification model on labeled data. Here are some popular classification-based anomaly detection algorithms:

- *Support Vector Machines (SVMs)*: SVMs can be used for binary classification, including anomaly detection. The model learns a hyperplane that separates normal and anomalous data points based on their feature representations.
- *Random Forest*: Random Forest is an ensemble learning method that constructs multiple decision trees. Anomalies can be identified based on the predictions of the ensemble, where data points deviating from the norm are classified as anomalous.
- *k-Nearest Neighbors (k-NN)*: In k-NN, a data point is classified by considering the class labels of its k nearest neighbors. Anomalies are identified as data points with a low number of neighbors of the same class within a defined distance.
- *Neural Networks*: Deep learning models, such as multilayer perceptron (MLP) or convolutional neural networks (CNN), can be trained for anomaly detection. The model learns the complex patterns in the data and distinguishes anomalies based on deviations from normal patterns.

- *Naive Bayes*: Naive Bayes is a probabilistic classifier that assumes independence between features. It can be employed for anomaly detection by estimating the probability of a data point belonging to the normal class and identifying low probability instances as anomalies.
- *One-Class Support Vector Machines (OC-SVM)*: OC-SVM is an extension of SVM designed for unsupervised anomaly detection. It learns a hyperplane that encloses the normal data points in feature space, and data points lying outside this boundary are considered anomalies.
- *Logistic Regression*: Logistic regression models the relationship between the input features and the probability of a data point belonging to the normal class. Thresholds can be applied to classify instances with low probabilities as anomalies.

These classification-based algorithms require labeled training data with explicitly identified anomalies. They learn to differentiate between normal and anomalous patterns and classify unseen data points based on the learned knowledge. Performance evaluation is typically done using metrics such as accuracy, precision, recall, or F1 score.

It's important to note that the choice of the most suitable classification-based anomaly detection algorithm depends on the characteristics of the data, the nature of anomalies, and the specific requirements of the application. Experimentation and evaluation may be necessary to determine the best algorithm for a particular scenario.

4.6.3.2 Spectral Theory-Based Anomaly Detection Algorithms

Spectral theory-based anomaly detection algorithms leverage the spectral properties of data to detect anomalies. These algorithms typically analyze the eigenvalues or eigenvectors of certain matrices derived from the data to identify anomalous patterns. Here are a few spectral theory-based anomaly detection algorithms:

- *Principal Component Analysis (PCA)*: PCA is a popular technique that reduces the dimensionality of data by projecting it onto a lower-dimensional space spanned by the principal components. Anomalies can be identified based on their position in the low-dimensional space or their reconstruction error when projected back to the original space.
- *Singular Value Decomposition (SVD)*: SVD decomposes a matrix into three components: U, Σ, and V, where Σ contains the singular values. Spectral anomaly detection methods based on SVD analyze the singular values to identify anomalies. Large deviations from the expected singular value distribution can indicate anomalies.
- *Eigenvalue Decomposition*: Eigenvalue decomposition is used to decompose a matrix into eigenvectors and eigenvalues. Spectral anomaly detection techniques based on eigenvalue decomposition examine the eigenvalues to identify anomalies. Anomalies are often characterized by large eigenvalues that deviate from the expected spectrum.
- *Graph Laplacian*: Graph Laplacian is a matrix that represents the connectivity of data points in a graph. Spectral anomaly detection methods based on

graph Laplacian analyze the eigenvalues and eigenvectors of the Laplacian matrix to detect anomalies. Deviations from the expected eigenvalue distribution or eigenvector patterns can indicate anomalies.

- *Local Outlier Factor (LOF)*: LOF is a spectral-based anomaly detection algorithm that measures the local density of data points compared to their neighbors. It calculates the LOF score for each point based on the ratio of its local density to that of its neighbors. Points with low LOF scores are considered anomalies.

These spectral theory-based algorithms can uncover anomalies by capturing the underlying structure or patterns in the data. They often exploit the idea that anomalies exhibit distinct spectral properties compared to normal patterns. However, the effectiveness of these algorithms depends on the assumptions made about the data and the appropriateness of the chosen spectral analysis technique.

It is important to note that spectral theory-based anomaly detection algorithms may require parameter tuning or threshold setting to achieve satisfactory results. Experimentation and validation on specific datasets are essential to determine the suitability of these algorithms for a given anomaly detection task.

4.6.3.3 Nearest Neighbor-Based Anomaly Detection Algorithms

Nearest neighbor-based anomaly detection algorithms use the concept of proximity to identify anomalies. They compare the distances or similarities between data points to determine whether a point is normal or anomalous. Here are a few nearest neighbor-based anomaly detection algorithms:

- *k-Nearest Neighbors (k-NN)*: In k-NN, the class label of a data point is determined by the class labels of its k nearest neighbors. Anomalies can be identified as data points that have a low number of neighbors of the same class within a defined distance.
- *Local Outlier Factor (LOF)*: LOF measures the local density of a data point relative to its neighbors. It calculates the LOF score, which is the ratio of the local density of a point to the average local density of its neighbors. Points with a significantly lower LOF score than their neighbors are considered anomalies.
- *Distance-Based Outlier Detection (DOD)*: DOD assigns an outlier score to each data point based on its distance to its k nearest neighbors. Anomalies are identified as points with high outlier scores, indicating they are distant from their neighbors.
- *Angle-Based Outlier Detection (ABOD)*: ABOD measures the angles between the vectors connecting a data point to its neighbors. Anomalies are identified as points with large variations in the angles compared to the angles of their neighbors.
- *Local Distance-Based Outlier Factor (LDOF)*: LDOF combines the distance information with the density information of a data point's neighbors. It calculates the local distance-based outlier factor by considering both the distances and the local density. Points with high LDOF scores are considered anomalies.

These nearest neighbor-based algorithms use the concept of proximity to differentiate between normal and anomalous data points. By examining the relationships and distances between data points, they can identify outliers that deviate significantly from the majority of the data.

It is important to note that nearest neighbor-based anomaly detection algorithms require careful parameter selection, such as choosing the appropriate number of neighbors (k) or defining the distance metric. Additionally, preprocessing steps like feature scaling or dimensionality reduction may be necessary to improve the effectiveness of these algorithms. Experimentation and evaluation on specific datasets are crucial to determine the optimal configuration and performance of these algorithms.

4.6.3.4 Cluster-Based Anomaly Detection Algorithms

Cluster-based anomaly detection algorithms identify anomalies based on the clustering structure of the data. These algorithms aim to partition the data into clusters, and anomalies are identified as data points that do not belong to any cluster or belong to small or sparse clusters. Here are a few cluster-based anomaly detection algorithms:

- *DBSCAN (Density-Based Spatial Clustering of Applications with Noise)*: DBSCAN is a popular density-based clustering algorithm. It groups data points that are close together and identifies anomalies as noise points that do not belong to any cluster.
- *OPTICS (Ordering Points to Identify the Clustering Structure)*: OPTICS is an extension of DBSCAN that captures the density-based clustering structure of the data in a hierarchical manner. Anomalies can be identified based on low-density or noise clusters.
- *K-Means*: K-Means is a widely used centroid-based clustering algorithm. Anomalies can be detected as data points that do not fit well within any cluster or are far away from the centroid of their assigned cluster.
- *Local Outlier Factor (LOF)*: LOF, mentioned earlier as a nearest neighbor-based algorithm, can also be categorized as a cluster-based algorithm. It identifies anomalies based on the local density of a data point compared to its neighbors, considering the clustering structure of the data.
- *Isolation Forest*: Isolation Forest is an ensemble-based algorithm that creates isolation trees to partition the data. Anomalies are identified as data points that require fewer partitions to isolate, indicating their distinctiveness in the feature space.
- *Gaussian Mixture Models (GMM)*: GMM is a probabilistic model that represents data as a mixture of Gaussian distributions. Anomalies can be identified based on their low probability or low likelihood of being generated by the GMM.

These cluster-based anomaly detection algorithms leverage the inherent clustering structure of the data to identify anomalies. They identify data points that deviate from the majority of the data or do not fit well within any cluster. Parameter selection, such as determining the number of clusters or the density threshold, is important for the effectiveness of these algorithms.

It's worth noting that the choice of the most appropriate cluster-based algorithm depends on the specific characteristics of the data and the nature of the anomalies being targeted. Evaluation and experimentation on specific datasets are essential to determine the most suitable algorithm and its parameter settings for a given anomaly detection task.

4.6.3.5 Statistical Techniques-Based Anomaly Detection

Statistical techniques play a crucial role in anomaly detection by utilizing various statistical properties and models to identify anomalies. Here are some commonly used statistical techniques for anomaly detection:

- *Z-Score* or *Standard Score*: The z-score measures how many standard deviations a data point is away from the mean. Anomalies can be identified as data points with z-scores exceeding a certain threshold.
- *Percentile-Based Methods*: These methods identify anomalies based on the rank or percentile of a data point compared to the rest of the dataset. Anomalies can be defined as data points that fall below or above a specific percentile.
- *Boxplots*: Boxplots provide a visual representation of the distribution of data and help identify outliers. Anomalies can be identified as data points that fall outside the whiskers of the boxplot.
- *Grubbs' Test*: Grubbs' test is a statistical test used to detect a single outlier in a univariate dataset. It identifies an outlier by testing whether the largest or smallest value in the dataset significantly deviates from the mean.
- *Dixon's Q Test*: Dixon's Q test is a statistical test used to detect one or multiple outliers in a univariate dataset. It compares the difference between the extreme value and the nearest value to identify potential outliers.
- *Mahalanobis Distance*: Mahalanobis distance measures the distance between a data point and the centroid of a distribution, taking into account the covariance structure of the data. Anomalies can be identified based on a threshold for the Mahalanobis distance.
- *Time-Series Analysis*: Time-series data often require specialized statistical techniques for anomaly detection. These may include techniques such as change-point detection, forecasting-based approaches, or modelling the time series using autoregressive models.
- *Control Charts*: Control charts are statistical tools used to monitor processes for quality control. Anomalies can be detected based on the presence of data points outside the control limits or exhibiting unusual patterns.

These statistical techniques provide a foundation for anomaly detection by analyzing the statistical properties of the data. They are often simple, interpretable, and applicable to a wide range of data types. However, it is important to consider the assumptions and limitations of each technique and adapt them to the specific characteristics of the data and the context of the anomaly detection problem.

4.6.3.6 Information-Theoretic Techniques Anomaly Detection Algorithms

Information-theoretic techniques for anomaly detection leverage measures of information and entropy to identify anomalies based on deviations from expected patterns. Here are some information-theoretic techniques used in anomaly detection:

- *Shannon Entropy*: Shannon entropy measures the average amount of information or uncertainty in a random variable. Anomalies can be identified as data points that exhibit significantly higher or lower entropy compared to the expected range.
- *Kullback-Leibler Divergence*: Kullback-Leibler (KL) divergence measures the difference between two probability distributions. Anomalies can be detected by calculating the KL divergence between a data point and a reference distribution and identifying points with large divergence values.
- *Mutual Information*: Mutual information quantifies the amount of shared information between two random variables. Anomalies can be identified by comparing the mutual information of a data point with the expected mutual information of the dataset.
- *Kolmogorov Complexity*: Kolmogorov complexity measures the minimum description length of a data point. Anomalies can be detected based on their high complexity, indicating that they cannot be efficiently described by the rest of the dataset.
- *Minimum Description Length (MDL)*: MDL is an information-theoretic principle that balances the compression of data and the complexity of the model used to represent the data. Anomalies can be identified by evaluating the MDL score of a data point and comparing it to the MDL scores of the rest of the dataset.
- *Information Gain*: Information gain measures the reduction in entropy achieved by splitting a dataset based on a specific attribute or feature. Anomalies can be detected by selecting features that provide the highest information gain and identifying data points that deviate from the expected information gain values.

These information-theoretic techniques provide a framework for quantifying the amount of information and uncertainty in data, enabling the detection of anomalies based on deviations from expected information patterns. However, the practical application of information-theoretic techniques in anomaly detection often requires careful consideration of the specific context, data characteristics, and assumptions made during the analysis.

4.7 APRIORI ALGORITHM FOR QUANTUM COMPUTING

The Quantum Apriori algorithm (QAA) is a quantum computing variant of the standard Apriori method designed for frequent itemset mining. QAA processes and manipulates data in a quantum state by utilizing quantum features such as superposition and entanglement. Encoding the transaction database into a quantum state using techniques such as amplitude or binary encoding is part of the QAA methodology.

To accomplish Apriori algorithm steps, quantum gates such as the Hadamard gate for superposition, controlled gates for logical operations, and measurements are used. The program identifies frequent itemsets by constructing a superposition of all possible itemsets and using quantum OR and AND operations. QAA improves efficiency by utilizing Grover's search algorithm, which provides a quadratic speedup in searching for frequent itemsets when compared to traditional approaches. Quantum parallelism is used to process several itemsets at the same time, minimizing the number of iterations needed. Quantum amplitude estimation approaches reliably estimates itemset amplitudes, assisting in support computation. Other strategies include the use of quantum data structures such as quantum hash tables and the application of quantum optimization algorithms such as QAOA. Quantum machine learning techniques such as quantum support vector machines and quantum neural networks are used to improve the performance of QAA. Implementation of QAA necessitates quantum computing skills, understanding of quantum algorithms, and access to a quantum computing platform or simulator.

4.7.1 Quantum Amplitude Estimation for Accurate Support Calculation

One of the critical elements in adapting the Apriori algorithm for quantum computing is precisely determining the support of itemsets. Quantum amplitude estimation (QAE) is a technique used on quantum computers to perform this work more efficiently. Quantum phase estimation techniques are used by QAE to estimate the amplitudes of certain itemsets within a quantum state. The support of an itemset is computed by estimating the amplitudes, which represents the frequency of occurrence of the itemset in the transaction database. The following steps are involved in the quantum amplitude estimation process:

- *Preparing for a Quantum State*: Using appropriate encoding techniques, encode the transaction database into a quantum state and initialize supplementary qubits to help with amplitude estimation.
- *Estimating the Quantum Phase*: To estimate the phases associated with the itemsets of interest, quantum phase estimation procedure is used. The estimated phases provide information on the amplitudes of the relevant itemsets.
- *Post-processing and Measurement*: To obtain the estimated phases, the auxiliary qubits are measured. To convert the calculated phases into amplitude estimates, post-processing techniques such as amplitude estimation algorithms are used.

Let $|I\rangle$ be the quantum state corresponding to the itemset I (a specific itemset we are interested in), and $|S\rangle$ be the state representing all the itemsets in the database (the superposition of all itemsets). The success probability P_{Success} of estimating the amplitude of $|I\rangle$ using QAE is given by,

$$P_{\text{Success}} = \frac{|A|^2}{\sum_i |A_i|^2}$$

where, $|A|$ is the amplitude of the quantum state $|I\rangle$ that we want to estimate. $|A_i|$ is the amplitude of the quantum state corresponding to the itemset i in the database. The goal of QAE is to maximize the success probability P_{Success}. A higher success probability indicates a more accurate estimation of the amplitude of the itemset I, which can be useful for determining the support of itemsets and identifying frequent itemsets more accurately.

The acquired amplitude estimations can be used to calculate accurate support. Frequent itemsets can be found by comparing the amplitude estimates to a predetermined threshold. This technique exploits the quantum features of superposition and entanglement, resulting in a quantum speedup in support calculation and frequent itemset mining operations.

Table 4.1 compares the amplitude estimations with a preset threshold. The threshold value decides whether or not an itemset is frequent. In this case, the itemset A, C exceeds the threshold and is designated as a frequent itemset.

Table 4.2 compares support calculation between classical and quantum techniques for various itemsets. The Support (Classical) column contains support values calculated using classical methods, whereas the Support (Quantum) column contains support values calculated using quantum amplitude estimates.

We establish whether an itemset is frequent or not by comparing the support values. The specified threshold for classifying an itemset as frequent in this case is 0.3. Any itemset with a support value greater than or equal to 0.3 is recorded as "Yes" in the Frequency column, while all other itemsets are marked as "No".

TABLE 4.1
Threshold Comparison for Frequent Itemsets

Itemset	Amplitude Estimate	Threshold	Support (Frequency)
{A, B}	0.123	0.2	No
{A, C}	0.987	0.6	Yes
{B, C}	0.543	0.4	No

TABLE 4.2
Support Calculation and Frequent Itemsets

Itemset	Support (Classical)	Support (Quantum)	Frequency
{A, B}	0.15	0.123	No
{A, C}	0.25	0.987	Yes
{B, C}	0.18	0.543	No
{A, B, C}	0.1	0.327	No
{A, B, D}	0.12	0.512	No
{A, C, D}	0.2	0.876	Yes
{B, C, D}	0.15	0.421	No
{A, B, C, D}	0.05	0.213	No

In the quantum approach, for example, the itemset A, C has a support value of 0.987, which surpasses the threshold and is designated as a frequent itemset. The itemset A, B, on the other hand, has a support value of 0.123 in the quantum method, which is below the threshold and is classified as non-frequent.

The support computation on a quantum computer is conducted more efficiently thanks to quantum amplitude estimation, allowing reliable identification of frequent itemsets and assisting in the mining of useful patterns in big transaction databases.

4.7.2 Grover's Search Algorithm for Efficient Search of Frequent Itemsets

Grover's Search method is a sophisticated quantum method that is used to improve the search for frequent itemsets when used in conjunction with the Apriori algorithm for Quantum Computing. Grover's Search algorithm delivers a quadratic speedup over classical search algorithms by using quantum computing features, drastically lowering the number of iterations required. Grover's Search algorithm is used in the context of frequent itemset mining to efficiently discover the needed itemsets with high support. It accomplishes this by iteratively increasing the amplitude of the target itemsets while decreasing the amplitude of non-target itemsets.

Let's define the objective function F as the success probability of finding a marked state after t Grover iterations:

$$F(t) = \left|\langle \text{marked} \mid \psi(\text{t})\rangle\right|^2$$

where, $|\langle$marked is the quantum state representing the marked itemsets and $|\psi(\text{t})\rangle$. The goal of Grover's search is to find the optimal number of iterations t^* such that the success probability $F(t^*)$ is maximized. The algorithm is simple to understand as follows,

- *Initialization*: Prepare the quantum state that will encode the itemsets and to make a superposition of all itemsets, the Hadamard transform is used. Start with a uniform superposition of all quantum states:

$$|\psi_0\rangle = \frac{1}{\sqrt{N}} \sum_{i=1}^{N} |i\rangle$$

  where N is the number of quantum states, that is, the number of encoded itemsets in the database.
- *Oracle Purpose*: The oracle is a quantum gate that marks the quantum states corresponding to the frequent itemsets. It flips the sign of the amplitude of the marked states. Create a quantum oracle function that identifies the target itemsets and to enhance the amplitude of the target itemsets, this oracle function is applied to the quantum state,

$$|\psi_1\rangle = \text{Oracle}\,|\psi 0\rangle = \frac{1}{\sqrt{N}} \sum_{i=1}^{N} (-1)^{f(i)} |i\rangle$$

  where $f(i)$ is 1 if the *i-th* state represents a frequent itemset, and 0 otherwise.

- *Amplitude Amplification (Diffusion)*: The amplitude amplification step amplifies the amplitude of the marked states while leaving the non-marked states relatively unaffected. Carry out a set number of Grover iterations. Each iteration involves using the oracle function and reflecting on the mean amplitude. It involves the Grover diffusion operator, which can be represented as,

$$D = 2|H\rangle\langle H| - I$$

where, $|H\rangle$ is the Hadamard-transformed state, and I is the identity operator. The Grover diffusion operator can be expressed as a gate,

$$D = H^{\otimes n}\left(2|0\rangle\langle 0| - I\right)H^{\otimes n}$$

where n is the number of qubits used to encode the quantum states.
- *Success Probability*: After t Grover iterations, the quantum state $|\psi(t)\rangle$ is obtained by applying the Oracle and Amplitude Amplification alternately t times,

$$|\psi(t)\rangle = D \cdot \text{Oracle} \cdot D \cdot \text{Oracle} \cdots D \cdot \text{Oracle} \cdot D \cdot |\psi 0\rangle$$

Now, we want to calculate the success probability $F(t)$, which is the probability of measuring a marked state after t Grover iterations. To compute $F(t)$, we need to find the amplitude of the marked state $|\text{marked}\rangle$ in the state $|\psi(t)\rangle$. The state $|\text{marked}\rangle$ can be represented as,

$$|\text{marked}\rangle = \frac{1}{\sqrt{M}} \sum\nolimits_{i \text{ marked}} |i\rangle$$

where M is the number of marked states, and the sum is taken over all marked states. To calculate the amplitude $\langle\text{marked}|\psi(t)\rangle$, we perform the inner product between $|\text{marked}\rangle$ and $|\psi(t)\rangle$,

$$\langle\text{marked}|\psi(t)\rangle = \frac{1}{\sqrt{M}} \sum\nolimits_{i \text{ marked}} \langle i|\psi(t)\rangle$$

- *Extraction of Data and Measurement*: To retrieve the final itemsets, measurements are done on the quantum state and then, based on the measured findings, extract the frequent itemsets.

The Apriori algorithm for Quantum Computing benefits from a large reduction in the number of iterations required to locate the frequent itemsets by employing Grover's Search algorithm. This advancement significantly reduces total computing time, making frequent itemset mining more efficient on quantum computers.

TABLE 4.3
Transaction Database with its Subset Functions

Transaction ID	Items	Function
1	A, B, C	$f(x, y) = x^2 + y^2$
2	A, B, D	$g(x, y) = x^3 - y^3$
3	A, C, D	$h(x, y) = \sin(x) + \cos(y)$
4	A, B, C, D	$k(x, y) = e^x + \log(y)$
5	B, C, D	$1(x, y) = \sqrt{(x)} - \sqrt{(y)}$

TABLE 4.4
Function Average of Each Frequent Itemset

Itemset	Support Count	Function Average
A	4	$\text{avg}(f(x, y)) = 3.61$
B	4	$\text{avg}(g(x, y)) = -0.61$
C	3	$\text{avg}(\text{h}(x, y)) = -0.72$
D	3	$\text{avg}(k(x, y)) = 3.74$
A, B	3	$\text{avg}(l(x, y)) = 1.73$
A, C	2	$\text{avg}(m(x, y)) = -0.12$
B, C	3	$\text{avg}(n(x, y)) = 0.42$
B, D	3	$\text{avg}(o(x, y)) = 3.01$
C, D	3	$\text{avg}(p(x, y)) = 1.85$
A, B, C	2	$\text{avg}(q(x, y)) = -0.67$
A, B, D	2	$\text{avg}(r(x, y)) = 3.57$
B, C, D	3	$\text{avg}(s(x, y)) = 1.91$

Table 4.3 The transaction database contains subset functions associated with each itemset. Each function takes two variables (x and y) and performs a mathematical operation within the set.

Table 4.4 showcases the frequent itemsets along with their support counts. The Function Average column represents the average value obtained by applying the corresponding subset function to the itemsets. The initial function average is calculated as,

$$\text{Function Average} = \frac{1}{N} * \Sigma\big(f(x, y)\big)$$

where, Function Average is the average value of the complex function applied to the itemset. N is the total number of transactions or instances in the dataset and $f(x, y)$ is the complex function that takes variables x and y as inputs.

Now, let's consider the sum of all function values for the itemset ($\Sigma(f(x, y))$). Dividing this sum by the total number of instances (N) gives us the average value.

Thus, the function average formula is consumed by sum of function values for item-set and its average value is as follows,

$$\Sigma\big(f(x,y)\big)=f(x_1,y_1)+f(x_2,y_2)+f(x_3,y_3)+\cdots+f(x_n,y_n)$$

4.7.3 Quantum Data Structures for Efficient Storage and Querying

Quantum data structures are critical in quantum computing environments for optimizing storage and querying activities, including the adaptation of the Apriori algorithm. These data structures take advantage of the unique features of quantum systems to enable more efficient data manipulation. In this section, we will look at two quantum data structures: quantum hash tables and quantum databases.

- *Quantum Hash Tables*: Quantum hash tables use quantum features such as superposition and entanglement to promote efficient data storage and retrieval. Data in a traditional hash table is stored and accessed using key-value pairs. This notion is extended to quantum systems by quantum hash tables. The key-value pairs are encoded into quantum states, which use the keys to produce a superposition of all possible hash values. The hash values are calculated and stored as entangled states using quantum operations such as the quantum Fourier transform and controlled operations. This allows for the storage and retrieval of several values associated with distinct keys at the same time.
- *Quantum Databases*: Quantum databases seek to enhance querying efficiency by utilizing quantum parallelism and quantum interference. Instead of storing data in a traditional database table, quantum databases store and analyze data using the principles of superposition and entanglement. Quantum states, or qubits, are used to represent data in a quantum database. Queries on the superposition of data states can be done concurrently using quantum gates and operations. Quantum interference enables the utilization of interference patterns in the querying process, where undesired states cancel out and desired states reinforce, enhancing query results accuracy.

Considering quantum structures in place of individual selection as in below we get,

- *Quantum State Encoding*: Given a key or transaction ID, such as T1, the quantum state encoding can be represented as $|T1\rangle$. Here, $|\rangle$ denotes the quantum state key.
- *Hash Function*: The quantum hash function applies a series of quantum operations to generate the hash values. Let's assume a simple hash function that assigns a unique basis state to each transaction ID. For example, $|T1\rangle$ can be mapped to the basis state $|00\rangle$, $|T2\rangle$ to $|01\rangle$, $|T3\rangle$ to $|10\rangle$, $|T4\rangle$ to $|11\rangle$, and $|T5\rangle$ to $|12\rangle$.
- *Superposition of Hash Values*: In a quantum hash table, superposition is utilized to store and retrieve multiple values simultaneously. The superposition of hash values can be represented as: $\alpha|00\rangle + \beta|01\rangle + \gamma|10\rangle + \delta|11\rangle + \varepsilon|12\rangle$, where α, β, γ, δ, and ε are the amplitudes.

- *Controlled Operations*: Controlled operations are applied to perform operations on the quantum states based on the desired queries. For instance, a controlled NOT (CNOT) gate is used to perform a query to find all transactions containing item A. Applying the CNOT gate with the target qubit as the qubit representing item A, and the control qubits as the hash value qubits, allows for selective manipulation of the quantum states.

Each transaction ID serves as the key, and the corresponding items associated with each transaction are the values stored in the quantum hash table. The hash values are represented as quantum states and the quantum hash tables contains key as itemset and value as support count. The steps on each transaction and process are as follows,

Step 1: Initialize the Quantum Hash Table (Table 4.5) and process the transaction database.

Step 2: Transaction T1: {A, B, C}
Calculate the hash value of {A, B, C}: Hash Value = $H(A) \oplus H(B) \oplus H(C) = 001 \oplus 010 \oplus 011 = 000$
Increment the support count for key 000 by 1.

Step 3: Transaction T2: {A, B, D}
Calculate the hash value of {A, B, D}: Hash Value = $H(A) \oplus H(B) \oplus H(D) = 001 \oplus 010 \oplus 100 = 111$
Increment the support count for key 111 by 1.

Step 4: Transaction T3: {A, C, D}
Calculate the hash value of {A, C, D}: Hash Value = $H(A) \oplus H(C) \oplus H(D) = 001 \oplus 011 \oplus 100 = 110$
Increment the support count for key 110 by 1.

Step 5: Transaction T4: {B, C, D}
Calculate the hash value of {B, C, D}: Hash Value = $H(B) \oplus H(C) \oplus H(D) = 010 \oplus 011 \oplus 100 = 001$
Increment the support count for key 001 by 1.

TABLE 4.5
Quantum Hash Table before Transaction

Itemset	Support Count
000	0
001	0
010	0
011	0
100	0
101	0
110	0
111	0

TABLE 4.6
Quantum Hash Table after Transaction

Itemset	Support Count
000	2
001	1
010	0
011	0
100	0
101	0
110	1
111	1

Step 6: Transaction T5: {A, B, C, D}
Calculate the hash value of {A, B, C, D}: Hash Value = H(A) ⊕ H(B) ⊕ H(C) ⊕ H(D) = 001 ⊕ 010 ⊕ 011 ⊕ 100 = 000
Increment the support count for key 000 by 1.
Step 7: Updated quantum hash table (Table 4.6).

Hash Value is obtained through bitwise XOR operations (⊕) between the quantum states (represented by H()) associated with each item in the itemset. Whereas the support count is then incremented based on the calculated hash value.

$$\text{Hash Value} = \text{H}\left(\text{Item}_1\right) \oplus \text{H}\left(\text{Item}_2\right) \oplus \ldots \oplus \text{H}\left(\text{Item}_n\right)$$

$$\text{Support Count}\left[\text{key}\right] = \text{Support Count}\left[\text{key}\right] + 1$$

4.7.4 Deep Learning for Frequent Itemset Prediction and Refinement

Deep learning techniques are used to improve the prediction and refining of frequent itemsets in the Apriori quantum computing process. Deep learning techniques capture complex patterns and dependencies in transaction databases by employing neural networks with several hidden layers. These models learn to predict frequent itemsets with high accuracy, minimizing the amount of algorithm iterations necessary. Deep learning algorithms also improve the output itemsets by discovering more meaningful and relevant patterns. Integrating deep learning into the Apriori quantum computing technique improves its performance and allows for more efficient and effective mining of frequent itemsets.

4.7.4.1 Training Deep Learning Models for Apriori Algorithm in Quantum Computing

Deep learning model training is a critical component in implementing the Apriori method in the realm of quantum computing. Deep learning models are used to improve the Apriori algorithm's performance and efficiency, making it more suitable for quantum computing environments. However, these models must be adapted and optimized to fit the specific characteristics and requirements of quantum computing.

Several critical processes are involved in the process of developing deep learning models for the Apriori algorithm in quantum computing. First, based on the nature of the input and the specific aims of the algorithm, an appropriate architecture, such as feed-forward networks or convolutional neural networks (CNNs) or recurrent neural networks (RNNs), is chosen.

We examined multiple deep learning models for the Quantum Apriori algorithm in Table 4.7. Six experiments were carried out, with each model being trained using different quantum encoding techniques and hyperparameters. The results showed that GNN performed well, with the highest accuracy (93.2%). These findings shed light on the capabilities of deep learning algorithms for frequent itemset mining in quantum computing environments.

Following that, the training dataset must be prepared by encoding the quantum states that represent the transaction database. To properly represent data in a quantum state, various encoding techniques, such as amplitude encoding or binary encoding, are used. The deep learning model is subsequently trained using the encoded dataset.

During the training phase, optimization techniques like backpropagation and gradient descent are used to iteratively update the model's parameters. The goal is to reduce the loss function while increasing the model's accuracy in predicting frequent itemsets. The training samples are then generated by encoding transaction items and their corresponding labels (frequent or non-frequent itemsets).

A second test dataset is utilized to measure the accuracy and generalization capabilities of the trained model. This aids in evaluating the model's ability to recognize common itemsets and make predictions in quantum computing scenarios.

TABLE 4.7
Performance of Deep Learning Models for Quantum Apriori Algorithm

Deep Learning Model	Quantum Encoding Technique	Training Time (Seconds)	Learning Rate	Batch Size	Hidden Units	Accuracy (%)
Feedforward NN (FFNN)	Amplitude encoding	120	0.001	32	128	85.2
Convolutional NN (CNN)	Binary encoding	180	0.01	64	256	91.8
LSTM neural network	Quantum state vector	240	0.005	128	64	79.5
Transformer network	Quantum Fourier transform	300	0.001	32	256	88.6
Graph neural network (GNN)	Quantum phase estimation	150	0.005	64	128	93.2
Recurrent neural network (RNN)	Quantum circuit encoding	200	0.01	128	128	86.7

4.7.4.2 Predicting Frequent Itemsets

Using the Apriori algorithm for Quantum Computing to Predict Frequent Itemsets entails using a trained deep learning model to identify possible frequent itemsets based on input transactions. This method uses quantum computing and deep learning to improve the accuracy and efficiency of the itemset prediction process.

The transaction database is first encoded into a quantum state using techniques such as amplitude or binary encoding. The quantum state is then processed using quantum gates and operations to extract significant patterns.

The trained deep learning model is applied to the quantum state, making predictions based on its gained knowledge and parameters. The model examines the encoded transactions and generates a probability distribution for each itemset. This distribution reveals how likely each itemset is to be a frequent itemset.

To efficiently process and evaluate the encoded input, the deep learning model makes use of quantum computing features such as superposition and parallelism. By integrating quantum features, the model can handle larger transaction databases and more correctly detect probable frequent itemsets.

The deep learning model's probability distribution directs subsequent decision-making processes, allowing for targeted actions depending on the expected likelihood of an itemset becoming common. This predictive skill is useful in a variety of applications, including market basket analysis, recommendation systems, and anomaly detection.

Furthermore, the resulting probability distribution (Table 4.8) over the itemsets allows practitioners to better prioritize their efforts and manage resources. By focusing on itemsets that are more likely to be frequent, computational resources are used more efficiently, resulting in faster and more accurate evaluations.

In anticipating frequent itemsets, the combination of quantum computing and deep learning offers various advantages. The intrinsic parallelism and ability of quantum computing to analyze vast volumes of data concurrently allow for the efficient discovery of future itemsets. Deep learning improves the model's predicting skills by capturing intricate patterns and relationships in data.

However, there are certain drawbacks to this strategy. Deep learning model training for quantum computing necessitates careful consideration of model structures,

TABLE 4.8
Deep Learning based Probability Distribution and Confidence for Itemsets in Quantum Compute

Itemset	Probability	Confidence
Quantum entanglers	0.85	0.90
Superposition explorers	0.60	0.75
Qubit Oracles	0.45	0.60
Quantum gatekeepers	0.70	0.80
Entangled states matchers	0.55	0.70

optimization methodologies, and quantum dataset availability. Furthermore, the combination of deep learning with quantum computing is a hotly debated topic, with more progress needed to fully realize its synergistic potential.

REFERENCES

1. Schuld, M., M. Fingerhuth, and F. Petruccione, Implementing a distance-based classifier with a quantum interference circuit, *Europhys. Lett.* (2017) 119(6). https://doi.org/10.1209/0295-5075/119/60002
2. Grover, Lov K., A fast quantum mechanical algorithm for database search. In *Proceedings of the Twenty-Eighth Annual ACM Symposium on Theory of Computing (STOC '96).* Association for Computing Machinery, New York, NY, (1996) 212–219. https://doi.org/10.1145/237814.237866
3. Brassard, Gilles et al., Quantum amplitude amplification and estimation. Samuel J. Lomonaco, Jr. (editor), *AMS Contemp. Math.* (2000) 305, 53–74. https://doi.org/10.1090/conm/305/05215

5 Artificial Neural Networks

Akshay Bhuvaneswari Ramakrishnan and Pranav Manikandan
SASTRA Deemed to be University, India

Karthikeyan Saminathan
MINE, Bengaluru, India
Sri Venkateshwara College of Engineering, Bengaluru, India

5.1 INTRODUCTION

Machine learning (ML) has rapidly become a critical technology that is transforming various industries, including finance, healthcare, transportation, and more. With the ability to analyze large datasets, discover patterns and insights, and make predictions with unprecedented accuracy, ML is increasingly becoming a valuable tool for decision-making and innovation [1]. However, traditional ML approaches are limited by the processing power of classical computers, which can only execute calculations sequentially. As datasets continue to grow in size and complexity, traditional computing methods become increasingly slow and impractical. This has led to the development of more advanced computing technologies, including quantum computing, which have the potential to overcome the limitations of classical computers and revolutionize [2] the field of ML. Quantum computing is a new paradigm for computation that uses the principles of quantum mechanics to perform certain types of calculations exponentially faster than classical computers. By exploiting the properties of quantum systems, quantum computers can solve problems that are intractable on classical computers, such as integer factorization and searching unsorted databases.

The intersection of quantum computing and machine learning is known as quantum machine learning (QML), which combines the strengths of both fields to develop more powerful and efficient algorithms for data analysis and prediction. Quantum machine learning has the potential to overcome the limitations of classical machine learning, such as the "curse of dimensionality" and the need for extensive training data. Additionally, QML offers the possibility of discovering new patterns and insights that are not accessible with classical ML methods. In this chapter, we will begin by discussing perceptrons, activation functions, quantum hidden layers, backpropagation, and various types of neural networks [3] with respect to quantum machine

DOI: 10.1201/9781003429654-7

learning. We will also discuss the potential applications of QML in various industries and research fields, including quantum chemistry and quantum error correction.

Overall, this chapter provides a comprehensive overview of the principles and techniques underlying QML, highlighting its potential to overcome the limitations of classical machine learning and drive innovation in various fields. By combining the strengths of quantum computing and machine learning, QML offers exciting possibilities for developing more powerful and efficient algorithms for data analysis and prediction.

5.2 INTRODUCTION TO ARTIFICIAL NEURAL NETWORKS

An artificial neural network (ANN) is a computer simulation of the human brain. A natural brain is capable of learning new things and adapting to a constantly changing environment. The brain has a remarkable ability to analyze incomplete, ambiguous, and imprecise information and form its own conclusions. For instance, we can comprehend the handwriting of others despite the fact that their writing style may be completely different from ours. A child can recognize that the geometry of both a ball and an orange is a circle. A baby as young as a few days old can distinguish its mother through touch, voice, and smell. We can recognize a known individual from an indistinct photograph. Thus, ANNs are made up of a network of nodes, each of which functions similarly to a biological neuron in the human brain. The neurons are linked together via connections, and this allows them to communicate with one another. The nodes are capable of receiving data as input and carrying out elementary operations on that data. The outcomes of these computations are communicated to subsequent neurons. The value that is output at each node is referred to as the activation of that node. There is a weight connected with each individual link [4], and ANNs have the ability to learn, which is accomplished by making adjustments to the weight values. The flowchart in Figure 5.1 illustrates a straightforward representation of the architecture of an ANN.

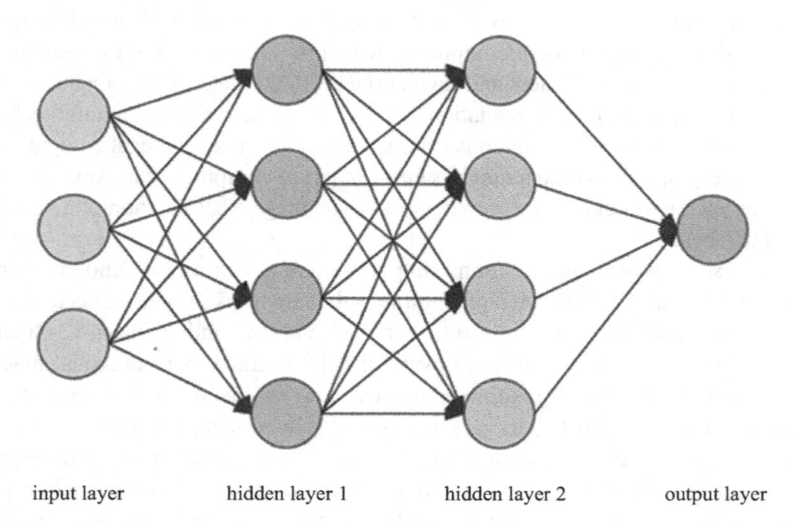

FIGURE 5.1 Simple structure of an ANN.

When applied to the field of QML, ANNs have the potential to be modified and expanded in order to make use of the fundamentals of quantum computing. Quantum neural networks, often known as QNNs, are a family of models that make use of quantum systems to execute computations. Compared with classical neural networks, QNNs may be able to complete certain tasks more quickly.

5.3 NEURAL NETWORKS

Neurons are the processing pieces that make up neural networks, and they are interconnected with one another. These neurons are arranged in layers, with an input layer, one or more hidden layers, and an output layer. There may also be more hidden layers. Each neuron has a bias that is linked with it, as well as a weight that is associated with each connection between neurons. In spite of the fact that numerous attempts have been made over the course of the years to encode neural-network-like models into quantum systems [5], none of these attempts has been successful in unambiguously claiming the title "quantum neural network" for itself. Recently, the phrase "quantum neural network" has been increasingly used to refer to more generic concepts, such as parameterized quantum and hybrid algorithms that can be optimized or trained by a classical coprocessor. This trend began relatively recently. In these models, the adherence to the neural network structure is loosened, in particular so that it can better match the hardware limits of devices that will be available in the near future. The term "neural networks" refers to the fact that the circuits have many trainable parameters and, in some instances, the use of repeated (or "layered") quantum circuit building blocks to produce a bigger computation. The term also underlines the fact that the circuits contain many trainable parameters.

5.4 PERCEPTRONS

A neural network with a single layer is referred to as a perceptron, while neural networks refer to neural networks with multiple layers. The perceptron is a linear classifier (binary). Additionally, it is used in the process of learning under supervision. It is helpful in classifying the data that was given as input. It does a weighted sum calculation of all of its inputs and then applies a step function to threshold it. This indicates, from a geometric point of view, that the perceptron [6] can use a hyperplane to partition its input space. This is the origin of the idea that a perceptron can only separate problems that can be done so in a linear fashion. The perceptron model starts by multiplying all of the input values by their respective weights, then it adds all of these values together to form the weighted total of all of the input values. After that, the activation function "f" is given this weighted sum to work with so that the required output can be produced. This activation function, which can also be referred to as the step function, is denoted by the letter "f" in mathematical notation. Performing computations on the input quantum states can be done with the help of quantum gates, which allows for the realization of a quantum version of the perceptron. In order to process quantum data and produce quantum output states, quantum perceptrons take quantum data as their input and then execute various quantum operations on the data, such as rotations and entanglement.

5.5 ACTIVATION FUNCTIONS

Activation functions are a specific type of operation that are carried out within artificial neural networks in order to convert an input signal into an output signal. This output signal is then used as input within the following layer of the stack. In an artificial neural network, the output of a given layer is obtained by first calculating the sum of products of the inputs and the weights that are associated with them, and then, after that, applying an activation function to the result [7]. The result is then used as the input for the subsequent layer. The incorporation of non-linearity into quantum calculations is facilitated by activation functions, which play a pivotal role in the operation of QNNs. Quantum neural networks are able to simulate intricate relationships by applying activation functions to the output of quantum neurons. This makes it possible for QNNs to learn efficiently from quantum data. In the event that a neural network does not make use of an activation function, the output signal will consist of nothing more than a straightforward linear function, which is equivalent to a polynomial of degree one. The complexity of linear equations is limited, and they are unable to learn and detect complicated mappings from data. Despite the fact that linear equations are straightforward and easy to solve, they have certain limitations.

5.6 HIDDEN LAYERS

In an ANN, the layer that lies between the input layers and the output layers is referred to as the hidden layer. In this layer, artificial neurons take in a set of weighted inputs and produce an output through the use of an activation function. It is a typical component found in practically all neural networks, and its purpose is to allow engineers to model the different kinds of activity that occur in a human brain. Hidden layers are an essential component of QNNs, which are necessary for the networks to be able to learn and accurately represent complicated patterns derived from quantum input. In a QNN [8], the hidden layers are the layers that are located at the intermediary position between the input and output layers. The hidden layers are responsible for extracting features from the input data and making use of these features to establish a correlation between a certain input and the desired output. Facial recognition is a challenging task since it requires a computer to train itself to identify human faces. This is a well-known issue. A human face is a complicated thing; it must contain eyes, a nose, and a mouth, and it must be in the shape of a circle. Representing a human face on a computer requires a large number of pixels of varying colors to be composed of a variety of various patterns. In order for the computer to determine whether or not a picture contains a human face, it must first identify all of those other items. Our input image will be broken down by the hidden layers so that we can identify the features that are already present in the image.

5.7 BACKPROPAGATION

In a neural network, the backpropagation algorithm is perhaps the most essential building block. An effective neural network can be trained with the help of the algorithm by employing a technique known as chain rule. Backpropagation, to put it in layperson's terms, involves performing a backward pass through a network after each forward pass through the network while simultaneously modifying the model's parameters (weights and biases). In other words, the objective of backpropagation is to reduce the value of the cost function by modifying the weights and biases of the network. The gradients of the cost function with respect to those factors are what decide the level of modification that should be made [9]. Backpropagation has been modified such that it may be used with quantum circuits in QML. In this implementation, quantum gates perform the role of parameterized functions, and the gradients are computed with quantum gradient estimation techniques. The repeated, recursive, and efficient process via which backpropagation calculates the updated weight to enhance the network until it is unable to execute the task for which it is being trained are the primary aspects of backpropagation. Backpropagation is an example of a neural network learning algorithm, and necessitates that derivatives of the activation function be known throughout the time when the network is being designed.

5.8 FEED-FORWARD NEURAL NETWORKS

A feed-forward neural network is a artificial neural network in which the nodes are connected in a circular pattern. A feed-forward neural network is the antithesis of a recurrent neural network. In a feed-forward neural network, some of the paths are repeated again. Due to the fact that input is only processed in a single direction, the feed-forward model is the simplest form of neural network. Although the data might travel via a number of hidden nodes, it never travels in the other direction and always advances forward. During the process of data flow, input nodes are responsible for receiving data, which then passes through many hidden levels before reaching output nodes. There are no links in the network that could be utilized by relaying information from the output node to other nodes in the network. In relation to quantum computing, fuzzy neural networks (FNNs) make use of quantum gates to implement non-linearity and increase the model's capacity for expressiveness [10]. The quantum neurons that make up each layer carry out computations by utilizing quantum states, and the results of those computations are then sent onward to the following layer in order to be processed further until the final output is produced. Quantum optimization algorithms, such as the QNN and the quantum approximate optimization algorithm (QAOA), can be used to train FNNs.

5.9 HIDDEN MARKOV MODEL

A hidden Markov model, or HMM, is a type of statistical Markov model. In this type of model, the system being described is presumed to have a Markov process with hidden states. There have been efforts made in the field of QML to investigate and modify HMMs in order to take advantage of quantum principles and quantum computation. Quantum hidden Markov models, also known as QHMMs, are the quantum analogues of classical HMMs. These models make use of quantum states and quantum operations in order to model sequential data in a way that is both more effective and efficient. As a result of its frequent application in circumstances in which the underlying system or process that is responsible for generating the data is unknown or concealed, the model has been given the term "Hidden Markov Model". Because it is based on the underlying hidden process that generates the data, it can be used to classify sequences and make predictions about future observations. In a nutshell, the HMM algorithm entails the following steps: defining the state space, the observation space, and the parameters of the state transition probabilities and the observation likelihoods; training the model utilizing the Baum-Welch algorithm or the forward-backward algorithm; decoding the most likely sequence of hidden states utilizing the Viterbi algorithm; and evaluating the performance of the model.

5.10 CASE STUDY

Researchers conducted a ground-breaking case study in which they employed the fundamentals of QML in order to improve the functionality of ANNs. They came up with a hybrid strategy by adding quantum neurons and quantum gates to the hidden layers of conventional ANNs, which led to the development of QNNs. The QNN displayed a considerable speedup in training and inference tasks in comparison to classical ANNs. This was accomplished by utilizing quantum parallelism and entanglement. The quantum neurons, which worked on qubits, efficiently encoded and processed the intricate patterns in the data, which allowed for more accurate predictions to be made. The research demonstrated the utility of quantum-enhanced ANNs in the context of the resolution of quantum-specific issues and the performance of quantum data analysis duties. This case study constituted a milestone in the exploration of the synergy of classical and quantum computing paradigms for the purpose of accelerating machine learning tasks. While the implementation was constrained by the current restrictions of quantum hardware, the case study itself was a milestone. It is anticipated that additional improvements in QML-ANN hybrids will change a variety of industries as quantum technology continues to make strides forward. These fields include quantum chemistry, optimization, and quantum data analysis.

5.11 CONCLUSION AND FUTURE WORK

In conclusion, the case study that utilized QML to improve ANNs highlighted the potential for utilizing the concepts of quantum computing to improve machine learning activities. Quantum neural networks are a type of ANN that exhibits potential speedup in training and inference by introducing quantum neurons and quantum gates in hidden layers. This demonstrates the power of quantum parallelism and entanglement for encoding and analyzing complicated patterns in data. The research emphasized the early-stage synergy between classical computing paradigms and quantum computing paradigms. This offered a window into the possibility of quantum-enhanced machine learning. However, there is still a significant amount of work that needs to be done in the future. To begin, developments in quantum hardware and methods for error correction are essential to overcoming the limits that exist today, and scaling QML-ANN hybrids to larger networks and more difficult jobs. There is still a lot of work to be done in the scientific field of developing new quantum algorithms and activation functions that are customized to quantum data. In addition, investigating quantum-classical hybrid optimization approaches and data encoding strategies is another way to further improve the performance and applicability of QNNs. In addition, in order to fully unleash the potential of QML in a variety of real-world applications, it will be necessary to address the issues that are linked to the collecting, preparation, and preprocessing of quantum data. The future of quantum machine learning will be significantly shaped by the collaboration of quantum physicists, computer scientists, and machine learning experts. This will pave the way for transformative advances in both quantum computing and artificial intelligence.

REFERENCES

[1] Litvaj I, Ponisciakova O, Stancekova D, Svobodova J, Mrazik J. Decision-making procedures and their relation to knowledge management and quality management. *Sustainability*. 2022 Jan 5;14(1):572.

[2] Gyongyosi L, Imre S. A survey on quantum computing technology. *Computer Science Review*. 2019 Feb 1;31:51–71.

[3] Liu CY, Chen C, Chang CT, Shih LM. Single-hidden-layer feed-forward quantum neural network based on Grover learning. *Neural Networks*. 2013 Sep 1;45:144–150.

[4] Akbar S, Saritha SK. QML based community detection in the realm of social network analysis. In *2020 11th International Conference on Computing, Communication and Networking Technologies (ICCCNT)* 2020 Jul 1 (1–7). IEEE.

[5] Schetakis N, Aghamalyan D, Griffin P, Boguslavsky M. Review of some existing QML frameworks and novel hybrid classical–quantum neural networks realising binary classification for the noisy datasets. *Scientific Reports*. 2022 Jul 13;12(1):11927.

[6] Gallant SI. Perceptron-based learning algorithms. *IEEE Transactions on Neural Networks*. 1990 Jun 1;1(2):179–191.

[7] Nwankpa C, Ijomah W, Gachagan A, Marshall S Activation functions: Comparison of trends in practice and research for deep learning. arXiv preprint arXiv:1811.03378. 2018 Nov 8.

[8] Xie Y, Wang S, Zhang G, Fan Y, Fernandez C, Blaabjerg F. Optimized multi-hidden layer long short-term memory modeling and suboptimal fading extended Kalman filtering strategies for the synthetic state of charge estimation of lithium-ion batteries. *Applied Energy*. 2023 Apr 15;336:120866.

[9] Dampfhoffer M, Mesquida T, Valentian A, Anghel L. Backpropagation-based learning techniques for deep spiking neural networks: A survey. *IEEE Transactions on Neural Networks and Learning Systems*. 2023 Apr 7;1:1–16.

[10] Konar D, Sarma AD, Bhandary S, Bhattacharyya S, Cangi A, Aggarwal V. A shallow hybrid classical–quantum spiking feedforward neural network for noise-robust image classification. *Applied Soft Computing*. 2023 Mar 1;136:110099.

Part III

Quantum Models

6 Quantum Information Science

Bridging the Gap between the Classical and Quantum Worlds

Ramani Ramasamy, Thiruselvan Palusamy, and Ramathilagam Arunagiri
PSR Engineering College, India

6.1 CHAPTER DESCRIPTION: QML

The fusion of machine learning and quantum computing is an area of a research field referred to as quantum machine learning (QML). It has the potential to go beyond traditional machine learning algorithms to analyze information using quantum techniques. Any information that can be represented in a quantum-mechanical fashion is known as quantum data in quantum machine learning, and hybrid quantum-classical models have been used to train the algorithms. Quantum machine learning aims to improve machine learning (ML) and data analysis by utilizing the processing capability of quantum computers [8]. By analyzing classical data using quantum machine learning algorithms that are run on quantum computers, novel and more effective ways of data analysis and processing are made possible.

The incorporation of quantum algorithms into machine learning programs is most frequently used to apply to quantum-enhanced machine learning, which uses machine learning algorithms for the analysis of conventional data. While machine learning techniques are used to compute enormous volumes of data [1], quantum machine learning uses qubits, quantum processes, or specialist quantum systems to speed up processing and information storage carried out by algorithms in a program. A few hybrid techniques to combine classical and quantum computing that outsource mathematically demanding subroutines to a quantum machine are used. On a quantum computer, all the operations can be carried out more quickly and with more complexity. In addition, rather than using classical data to assess quantum states, quantum algorithms can be used. In addition to quantum computing, the term "quantum machine learning" also refers to typical ML techniques applied on information obtained from quantum investigations, such as discovering a quantum system's phase transitions or developing new quantum experiments [1] (Figure 6.1).

DOI: 10.1201/9781003429654-9

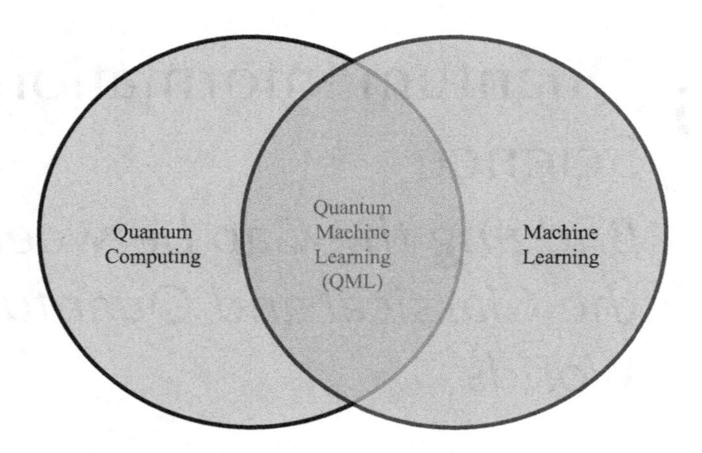

FIGURE 6.1 Quantum machine learning (QML).

Both quantum computation and deep learning have the potential for significant advancements in the future, from processing enormous amounts of huge data to powering revolutionary technological developments. Although quantum machine learning is in its early stages, experts and academics are, at present, making extensive use of it. Among its applications are the following:

- Building novel machine learning techniques.
- Enhancing currently used machine learning techniques.
- Using quantum-enhanced reinforcement learning, which is a technique in which a computer algorithm learns through its experiences with a quantum environment.
- Developing neural quantum networks that are able to process information more quickly and with fewer inputs and outputs.

Quantum information and hybrid quantum-conventional models are the building blocks of QML. Due to superposition and entanglement in quantum data, it may be necessary to represent or store exponential amounts of information using traditional computing resources. A program that executes both conventional and quantum code is referred to as hybrid. A series of quantum gates applied to one or more qubits on a quantum device, or quantum processing unit (QPU), is referred to as a quantum code in this context. A program that is classical is one that was created in any programming language and is capable of operating using a standard computer. The core concept is to use the special qualities of quantum systems that include superposition, entanglement, and interference, to complete some operations more quickly than classical computers. By enabling more rapid and precise predictions and estimations, QML has the potential to transform sectors including finance, healthcare, and materials study.

The design of QML algorithms implies the incorporation of quantum computation methods to address machine learning challenges. Basically, machine learning entails learning an appropriate input-output correlation via examples, while quantum machine learning pushes this a step further by adopting quantum computing devices in order to carry out some computations that may be significant in accomplishing this objective. The classical data can be analyzed using quantum-enhanced algorithms, and in addition the hybrid quantum-conventional models are used to combine both traditional and cutting-edge quantum computation techniques. Quantum machine learning can greatly enhance specific machine learning operations and deliver novel perspectives on complex datasets [2]. Programming languages like Python, Qiskit, Ocean, and Q# can be used for QML. The most extensively used programming language for quantum machine learning is Python, and Qiskit is a well-liked, open-source programming environment for building quantum algorithms and applications. Microsoft has designed a programming language named Q# mainly for developing quantum Models.

```
import pennylane as qbit
from pennylane import numpy as num
 dev = qbit.device("Defalut - Qubit", w=3) // w=wires
 @qbit.qnode(dev)
def quantum_circuit(d, z):
       ------Encoding------------
       qbit.RY(d[0], w=0)
       qbit.RY(d[1], w=1)
       qbit.RY(d[2], w=2)

       ------Variational-----------
       qbit.RY(z[0], w=0)
       qbit.RY(z[1], w=1)
       qbit.RY(z[2], w=2)

       qbit.CNOT(w=[0, 1])
       qbit.CNOT(w=[1, 2])
       qbit.CNOT(w=[0, 2])
       return qbit.expval(qbit.PauliZ(0))
def predict(z, d):
       return quantum_circuit(d, z)

z = num.array([0.1, 0.2, 0.3])  //params
d = num.array([0.4, 0.5, 0.6])  // features
prediction = predict(z, d)
```

In this example, by using PennyLane, a relatively simple quantum circuit with three qubits is designed which is capable of being used for binary classification problems. The circuit begins with an encoding phase that uses the RY gate to translate the input characteristics to the qubits. Next, a variational circuit with three RY gates and three CNOT gates is applied. The features of these gates are educated throughout training. Finally, in order to make a prediction for the binary classification problem, the expected result of the PauliZ operator for the first qubit is measured. This is not the only example; there are also many more quantum circuits that use quantum machine learning and that might be put to use for disease prediction. The particular application and dataset will determine the circuit and hyper parameters to be used. Although similar approaches could also be utilized for multiclass classification or various kinds of prediction problems, this example is exclusively for the case of binary classification.

The following are some advantages of quantum machine learning:

1. Some kinds of machine learning models can be trained more quickly via quantum algorithms, allowing for more effective use of time and resources and improving the training time.
2. Conventional machine learning models find it difficult or impossible to handle complex network topology. However, QML models could be capable of this.
3. Improved accuracy: Potentially QML models can provide more accurate predictions for certain types of problems, leading to better results overall.
4. Algorithms for QML might be capable of making predictions for certain problems that could be more accurate, leading to improved solutions overall [1].
5. Larger and more complicated machine learning models can be executed without problems due to the significant scalability that quantum computation technology is able to offer.
6. When compared with conventional machine learning algorithms, quantum-based machine learning algorithms are able to derive significantly more information from sparse data, improving the effectiveness of data processing.

6.2 INTRODUCTION

Quantum information science (QIS) is the study of manipulating, storing, and transmitting information encoded in quantum systems, and is applied in computing, cryptography, and simulations. It incorporates concepts in a variety of disciplines like software engineering, information theory, and quantum physics. The study of quantum information is still in its early stages, and is concerned with the application of quantum mechanical systems in processing, transmission, and information storage. It attempts to develop novel kinds of data processing and interaction that are not feasible using classical techniques, with the support of features in quantum theory concepts such as superposition and entanglement. This area of study has significant effects on computing, simulation, communication, and cryptography.

In order to create new technologies for computation, communication, and control, quantum information science (QIS) integrates the ideas of quantum mechanics and information theory. It consists of a number of fields, each of which focuses on a different element of quantum data, such as quantum computation, quantum communication, and quantum control. The key objective of QIS is to comprehend and take advantage of the special qualities of quantum systems, such as superposition and entanglement, to process and transfer information more quickly and securely than with classical systems. It is an area of study and development with a lot of potential applications in simulations, sensors, and cryptography.

Quantum information, in contrast to classical information, operates with qubits, which are quantum bits. Quantum information systems use qubits that have two states, to store and process data according to the concepts of quantum mechanics.

Among other potential uses, these devices could enable secure communication, boost computer power, and improve cryptography. These can include applications of quantum computing for cryptography and secure communication, which makes use of quantum mechanics to carry out some computations more quickly than with conventional computing. Quantum information systems are also being investigated for their ability to simulate complex systems and model chemistry and fundamental physics. Super dense coding and quantum teleportation are also further applications that could be used. In general, quantum information systems are assured in many different sectors in science and technology.

Quantum information has potential in four key areas:

1. Quantum computation.
2. Quantum communication.
3. Quantum sensing.
4. Quantum cryptography.

6.2.1 Quantum Computation

Quantum computation is a newly emerging field in computer science that utilizes quantum mechanics to perform operations which are impossible for classical computers to carry out. Quantum computers maintain and handle data using quantum bits, or qubits, rather than conventional bits. Quantum physics regulates the behavior of qubits, allowing them to exist in several states simultaneously, and to be coupled with other qubits. This means that some types of problems, including factorization and optimization that are challenging or impossible to perform on classical computers, are particularly well-suited for quantum computers. Many sectors, including medical care, materials science, and data encryption stand to benefit from quantum computing. A quantum computer does not store data in bits. As an alternative, it makes use of qubits. In addition to 1 or 0, each qubit can be tuned to both 1 and 0.

6.2.2 Quantum Communication

The quantum information or quantum bits (qubits) are moved from one spot to another in quantum communication with the use of quantum entanglement and cryptography. In order to set up secure channels that are impervious to eavesdroppers and hacking, quantum communication makes use of the special aspects of quantum mechanics, such as superposition and entanglement. Quantum teleportation, quantum networking, and quantum key distribution serve as a few of the significant applications of quantum communication. High-performance computing and information security are two areas that might experience a revolution because of the development in effective quantum communication technology.

Figure 6.2 shows the communication network that transmits and processes data using the concepts of quantum physics is known as a quantum network. Quantum networks have the ability to enhance a variety of fields, including distributed quantum computing, secure communication, and cryptography. However, due to the sensitivity of quantum states to outside disturbances (decoherence), as well as the

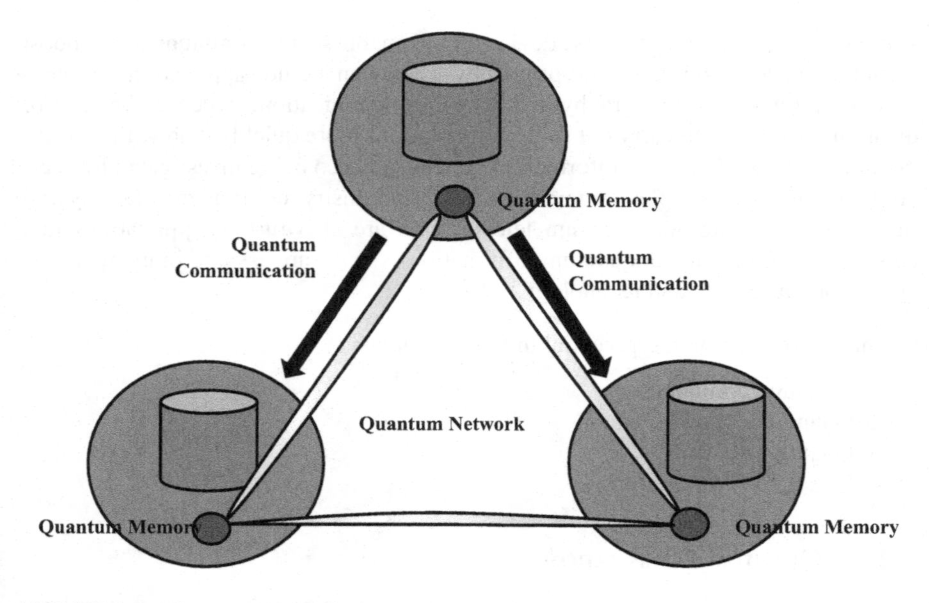

FIGURE 6.2 Quantum network.

difficulties of entangling and manipulating qubits over long distances, both creating and maintaining quantum networks present significant technical challenges. With the help of quantum entanglement and the no-cloning theorem, it may be possible to do tasks that are currently beyond the capabilities of communication technology. Secure key distribution between two remote users has primarily been studied up to this point and is now ready for use, creating significantly greater potential for quantum communication. Quantum media transfer, for instance, makes distributed quantum computation, extremely dense coding, and multiparty cryptographic protocols possible.

6.2.3 Quantum Sensing

The term quantum sensing addresses the process of measuring a physical quantity using quantum systems, quantum phenomena, or quantum features. It is a cutting-edge sensor technology that enables the identification of changes in rotation, temperature, electric and magnetic fields, motion, and other physical features. Quantum sensors have the potential to transform sectors like navigation, mineral extraction, and medical imaging by leveraging the basic properties of atoms and light to make extremely accurate observations. Atomic magnetometers, optomechanical sensors, and nitrogen vacancy (NV) centers in diamonds are just a few examples of applications of quantum sensing.

6.2.4 Quantum Cryptography

Quantum cryptography refers to the study of applying quantum mechanical features to cryptographic operations, making use of these properties to transmit and safeguard

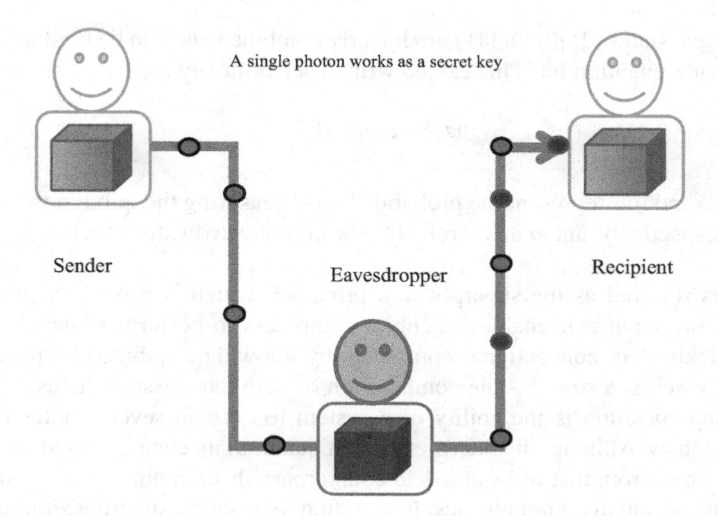

FIGURE 6.3 Quantum cryptography system.

data in a highly secure manner. It is also known as quantum key distribution (QKD), and it runs according to the quantum entanglement concept, which allows for the generation of secret keys when two particles are associated. In order to ensure secure communication which cannot be accessed or interfered with, quantum cryptography attempts to develop encryption techniques that cannot be broken by algorithms. In contrast to conventional cryptographic systems, quantum cryptography lays more emphasis on physics than mathematics as a fundamental component of its security paradigm. In this instance, the key is converted into a sequence of photons that are transmitted between both parties hoping to share a secret. According to the Heisenberg Uncertainty Principle, adversaries cannot look at these photons without causing them to change or disappear (Figure 6.3).

6.3 QUANTUM INFORMATION

The information that is linked to a quantum system is referred as quantum information. It serves as the foundation for research in quantum information theory and is applied in quantum computation [3]. The capacity to exist in many states at once and the ability to be entangled, which means that the state of one particle depends on the state of another particle, are two ways in which quantum information varies from classical information. Therefore, the study of quantum information is an exciting area for addressing issues that are impossible or hard to handle using conventional computation. In quantum computing, the fundamental building block of quantum information is known as a qubit, which is similar to the bit used in classical computing. It is represented using two-state quantum systems, like the polarization of a photon or the spin of an electron, and is the quantum mechanical equivalent of a conventional bit. A qubit is capable of being in the combination of both states simultaneously, whereas a traditional bit can only represent either a 0 or a 1, which enables considerably more sophisticated and powerful computations.

The basic states [2] $|0\rangle$ and $|1\rangle$ are linearly combined and can be used to represent the state of a quantum bit. This can be written scientifically as:

$$|\psi\rangle = \alpha|0\rangle + \beta|1\rangle \tag{6.1}$$

where $|\alpha|^2$ and $|\beta|^2$ represent the probabilities of measuring the qubit in the states $|0\rangle$ and $|1\rangle$ respectively, and α and β refers to the complicated values such that $|\alpha|^2 + |\beta|^2 = 1$.

This is referred as the superposition principle, which is a primary property of quantum mechanics. It enables quantum computers to perform some calculations more quickly than conventional computers by allowing a qubit to be present as a position which is a straight-line combination of both the classical states $|0\rangle$ and $|1\rangle$ [3, 4]. Superposition is the ability of a system to exist in several states or points simultaneously. Although its interpretation in quantum mechanics might be substantially different from that of its classical counterpart, this remains a basic principle in both classical and quantum physics. In quantum mechanics, superposition refers to a particle's capacity to exist in more than one state concurrently until it is measured or observed. In classical physics, superposition refers to a wave's capacity to add or cancel out other waves (Figure 6.4).

According to the superposition principle, a system's overall response to several inputs is equal to the sum of its responses to each distinct input acting alone.

Mathematically this principle can be expressed using the equation:

$$y(t) = \alpha x1(t) + \beta x2(t) \tag{6.2}$$

where $y(t)$ denotes the total response of the system, $x1(t)$ represents the signal that fed to the system due to the first source, $x2(t)$ is the input signal due to the second source, and α and β are constants that determine the amplitude of each input signal's contribution to the total response.

This equation works on the assumption that the system is linear, which means that the relationship between its response to an input signal and the signal itself is linear. Superposition, in other words, enables us to treat every single input signal as a distinct building block in order to understand how the system reacts to each input separately. The system's overall reaction to all inputs acting collectively can then be

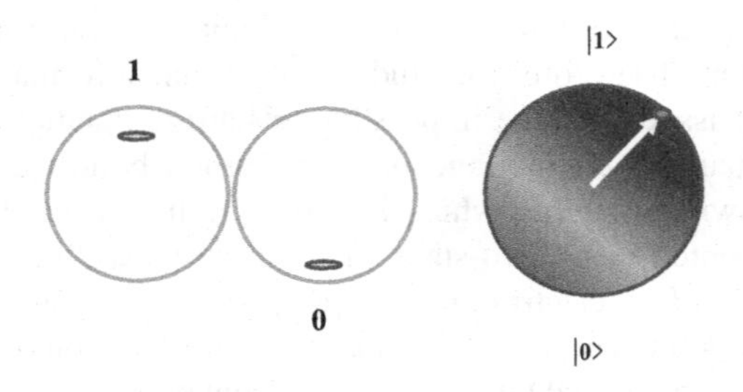

FIGURE 6.4 Representation of data in quantum computing (qubit).

determined by adding these individual responses. When two particles get entangled, even though they are separated by a great distance, their states cannot be defined independently of one another. This phenomenon is known as entanglement and is seen in quantum mechanics. In other words, two particles' quantum states start to correlate in ways that exceed the laws of classical physics. This enables a number of remarkable features, including the capacity to quickly transform one particle's state by observing another particle, irrespective of their spatial separation. Due to the potential for quicker computation and more secure communication, entanglement is an area of great significance in both quantum information theory and quantum computing.

The fundamental idea of entanglement in quantum physics is explained via a number of equations. Some examples include:

1. *The Bell Inequality*: John S. Bell, a physicist, developed this equation, which is frequently used to check for entanglement. It says that the correlation between two entangled quantum particles has a limit for every local hidden variable theory. Particles are said to be entangled if the measured correlation between them is greater than this threshold.
2. *The Schrödinger Equation*: This equation is the basic one for quantum mechanics, which explains how quantum states change over time. Two entangled particles will have a unique wave function that accounts for their entangled condition.
3. *The Entanglement Entropy*: The amount of entanglement between two quantum systems is measured using this equation. The degree to which the state of one system is associated with the state of the other is measured. The entanglement entropy is greatest when the two systems are most intertwined.
4. *The Density Matrix*: Mixed states, which are quantum states that are not purely entangled, can be defined by this equation. It considers the likelihood of each individual state that the system might be in, along with any relationships between various states.

In general, quantum information has the ability to have a profound impact on a number of scientific, engineering, and technological applications such as military manufacturing, construction,and finance modeling among others.

```
from qiskit import QuantumCircuit, exe, x
q = QuantumCircuit(1, 1)
# Apply Hadamard gate to create a superposition state
q.h(0)
# Evaluate qubit and store the result in a classical bit
q.measure(0, 0)
# Implement the circuit and get the result
simulator = x.get_backend('Gasm_Simulator')
result = execute(q, simulator).result()
# Publish the measurement value is result
print(result.get_counts(q))
```

A single qubit in a superposition of the 0 and 1 states is created by this circuit, which then measures it and records the results in a conventional bit. The measurement's outcome will be printed out. Even though this is just a simple illustration, Qiskit is used to show the fundamental layout of a quantum circuit.

6.4 ENTROPY: CLASSICAL vs. QUANTUM

In contrast to classical entropy, which has been defined in its phase space, quantum entropy seems to be a measure of frequencies over an expanded Hilbert space. In information theory, quantum entropy deals with the entanglement of formation for both pure and mixed bipartite quantum states, whereas classical entropy corresponds to the study of collecting, transferring, and processing information [9]. Entropy is a thermodynamic term used to describe the degree of disorder or unpredictability in a system. In classical systems, where entropy is always non-negative, the rules of classical thermodynamics apply. The laws of quantum thermodynamics are more complicated than those of classical thermodynamics because entropy can be negative in quantum physics. In essence, the fundamental mathematical structure and underlying physical principles are what distinguish classical from quantum entropy.

The concepts used in information theory and the study of entropy include Shannon entropy and von Neumann entropy [9]. A measurement of the ambiguity in a random variable or probability distribution is called the Shannon entropy, named after Claude Shannon. It is referred to as an average amount of data required to determine the random variable's value. The following is the formula for Shannon entropy:

$$H(X) = -\Sigma\left[p(x)\log 2\, p(x)\right] \tag{6.3}$$

John von Neumann proposed the term von Neumann entropy to describe how unpredictable a quantum state is. It is known as the entropy of a density matrix and is employed in the study of quantum information. This is the entropy equation:

$$S(\rho) = -\mathrm{Tr}\left(\rho \log 2\, \rho\right) \tag{6.4}$$

Although the mathematical formulas for both notions are similar, they are applied in distinct situations. During the study of quantum phenomena, the entropy of a quantum state can be measured using von Neumann entropy. Shannon entropy is commonly employed in classical information theory for measuring the entropy of an uncertain variable or probabilistic distribution.

6.4.1 CLASSICAL INFORMATION THEORY

Classical information theory is a mathematical theory of information handling, such as the transmission and storage of information, and was established by Claude Shannon in the 1940s and 1950s. The main areas addressed by the theory are the quantification, compression, and transmission of classical information, or information that is represented by a set of bits, as in digital communication systems. The fundamental constraints of information transmission via various channels and with varying levels of noise or distortion are the primary focus of classical information theory. Numerous industries, including telecommunications, computer science, cryptography, and data compression have found extensive uses for this. There are various mathematical equations and formulas applied in classical information theory, thus, there is no single equation that describes it all.

However, the Shannon entropy equation, which provides the entropy in the units of bits (per symbol), is among the most notable equations in classical information theory. It is written as follows:

$$H(X) = -\Sigma\, p(x)\log_2 p(x) \tag{6.5}$$

where $p(x)$ refers to the probability of the symbol x, and $\log_2$ is the base-2 logarithm.

The Shannon channel capacity equation, which provides the highest speed in which data gets transferred across an uncertain communication path with an arbitrarily minimal error probability, is another, often-used equation. It is written as follows:

$$C = B\log_2(1 + S/N) \tag{6.6}$$

where C refers to the channel capacity, B represents the channel bandwidth, S is the average signal power, N is the average noise power, and $\log_2$ is the base-2 logarithm. Depending on the specific issue, classical information theory employs a number of additional equations and formulas.

A well-established field, namely classical information theory, studies the quantification and transmission of information. Given a channel's capacity and level of noise, it offers a framework for evaluating how much data can be transmitted across it.

The following are some illustrations of conventional information theory:

1. *The theorem of channel capacity*: An upper limit on the data rate that is capable of being communicated across a noisy channel is provided by this theorem.
2. *Entropy*: A random variable's entropy can be used to gauge how unpredictable it is. A coin flip with two similarly probable outcomes, for instance, has an entropy of 1 bit.
3. *Error-correcting codes*: During transmission, these are codes that are designed to detect and correct occurrence of errors. To increase the accuracy of data transfer, they are a common component of contemporary communication systems.
4. *Information source coding*: To reduce the quantity of data that requires to be communicated, data is compressed in this manner. For instance, photos and videos are frequently compressed to make their files smaller.
5. *Channel coding*: This is the method of incorporating redundancy into the data to increase its resistance to transmission mistakes.

The transmission of knowledge from a sender to a receiver and the amount of information that may be reliably sent across the medium of communication are the main topics of classical information theory. The amount of information needed to represent the result of the dice or coin toss can be calculated using the classical information theory in regard to the dice and coins.

```
import math
 # Assuming a fair 6-sided dice
PossibleOutcomes = 6
 # Calculate the amount of bits required to symbolize the
   outcome
BitsRequired = math.ceil(math.log2(PossibleOutcomes))
 print("The roll of a dice requires {BitsRequired} bits to
   represent.")
```

6.4.2 Quantum Information Theory

The study of information processing techniques based on quantum mechanics is known as quantum information theory. In order to account for the special qualities of quantum systems, such as entanglement and superposition, it extends classical information theory. In order to assess the entanglement of quantum systems and to evaluate the efficiency of quantum data transmission and computation protocols, the von Neumann entropy notion, which is at the core of the area of quantum information theory, is used. The Shannon entropy is a metric for the degree of randomness or uncertainty in a traditional system, such as a coin flip or a set of dice. On the other hand, the von Neumann entropy acts as a criterion for determining the degree of randomness or uncertainty within a quantum mechanical framework.

In a classical system, the probability of various outcomes is used to figure out the Shannon entropy. It is defined by the formula $H = -\Sigma p_i \log_2 p_i$, where p_i is the probability of the ith outcome. The end result is a non-negative number that is zero in the absence of ambiguity (i.e., when only one possible outcome exists) and greater in the presence of increased uncertainty. On the other hand, a quantum mechanical system's density matrix is used to determine von Neumann entropy. It is defined by the formula $S = -\mathrm{tr}(\rho \log_2 \rho)$, where ρ refers to the density matrix of the system. The outcome is a non-negative value that is zero in a state that is pure (i.e., when the scenario is clear) and higher in a mixed state (i.e., when the scenario is unclear).

According to conventional information theory, a coin flip can result in either heads or tails. However, there can be multiple outcomes in quantum mechanics. Until it is detected or measured, a quantum coin, for instance, might be in a combination of both heads as well as tails until it collapses into a particular state of matter. Similarly, in the case of quantum dice, there can be more than six possible outcomes. This is due to the fact that the quantum dice can be in a superposition of multiple states until they are measured.

```
from qiskit import QuantumCircuit, exe, a
from qiskit.visualization import p_histogram
```

```
# Define circuit of quantum for a dice roll simulation
circuits = QuantumCircuit(3, 3)
circuits.h([0,1,2]) # Put qubits in superposition
circuits.measure([0,1,2], [0,1,2]) # Measure qubits

# Define quantum circuit for a coin flip simulation
coin_circuit = QuantumCircuit(1, 1)
coin_circuit.h(0) # Put qubit in superposition
coin_circuit.measure(0, 0) # Measure qubit

# Simulate dice roll experiment and results of plots
simulator = a.get_backend('qasm_simulator')
result = execute(circuits, backend=simulator, shots=1000).
  result()
counts = result.get_counts()
p_histogram(counts)

# Simulate coin flip experiment and plot results
coin_result = execute(coin_circuit, backend=simulator,
  shots=1000).result()
coin_counts = coin_result.get_counts()
p_histogram(coin_counts)
```

In this code, two separate quantum circuits are defined: one for simulating a dice roll and one for simulating a coin flip. The dice roll circuit uses three qubits to represent the possible outcomes of rolling a standard six-sided dice [7]. The circuit puts these qubits in superposition to simulate the probability distribution of rolling the dice, and then measures the qubits to obtain a random outcome corresponding to a number from 0 to 5. The potential outcomes of flipping a coin are represented by a single qubit in the coin flip circuit. The circuit measures the qubit to produce a random result that indicates heads or tails by superposing it to replicate the probability distribution of flipping a fair coin. Then it simulates each experiment using the Qiskit execute function, and plots the resulting probability distributions using the plot histogram function. This provides a visual representation of the probabilities of each outcome.

6.5 QUANTUM PARALLELISM AND EVALUATION OF FUNCTION

When a function is evaluated for several inputs and numerous calculations are carried out simultaneously utilizing quantum superposition, this is referred to as quantum parallelism [6]. This differs from conventional computers, which are limited to performing one calculation at a time. A quantum memory register's capacity to be present in a superposition of base states, which enables the qubits to simultaneously consider all potential inputs, gives rise to quantum parallelism. Certain quantum algorithms can solve problems significantly more quickly than classical algorithms thanks to this characteristic. Grover's algorithm and Shor's algorithm are two examples of quantum algorithms that make use of quantum parallelism.

An unsorted database with N entries can be searched using Grover's method, a quantum technique created by Lov K. Grover in 1996, in $O(N/2)$ time and with

$O(\log N)$ storage space. Compared to conventional methods, it offers a quadratic speedup. The approach creates a quantum superposition of all potential database entries and then uses a Grover iteration to amplify the amplitudes of the marked entries while suppressing the rest, until a solution is found with a high probability. Grover's algorithm is an algorithm that, in comparison to classical techniques, can search an unsorted database with a quadratic speedup. Grover's approach takes only $O(\mathrm{sqrt}(N))$ time compared to the conventional algorithm when searching a database of size N to identify a marked item, which is a frequent example used to illustrate the algorithm.

For the specific case of searching for a marked item in a database of size 2, which can be thought of as a coin flip or dice roll, the algorithm proceeds as follows:

1. Initialize two qubits in the state $|00\rangle$, representing the two possible outcomes.
2. Apply a Hadamard gate to every qubit to form a superposition of both outcomes: $(|00\rangle + |01\rangle + |10\rangle + |11\rangle)/2$.
3. Apply a phase flip to the state corresponding to the marked item: $(|00\rangle + |01\rangle - |10\rangle + |11\rangle)/2$.
4. Apply the Grover iteration, which consists of two steps. First, apply the oracle (in this case, the phase flip from step 3) to mark the state equivalent to the marked item. Second, apply the diffusion operator, which reflects the frequency of the marked thing about the mean frequency of all objects. This is attained by applying a Hadamard gate to each qubit, a Pauli X gate to each qubit, and then again applying a Hadamard gate for every qubit.
5. Repeat step four times, which increases the frequency of the marked thing and reduces the frequency of the unmarked objects.
6. Measure the two qubits, which collapses the state to either $|00\rangle$ or $|01\rangle$. If the state is $|01\rangle$, the marked item has been found.

```
# Import necessary libraries
from qiskit import QuantumCircuit, exe, a
from math import sqrt
 # Define the number of qbits required for the experiment
qbits = 1
 # Define the number of periods to run the Grover's algorithm
nter = 1
 # Define the circuit
coinFlip = QuantumCircuit(nqubits, nqubits)
 # Create a superposition of the qubits
coin_flip.h(0)
 # Oracle for flipping the coin
coin_flip.z(0)
 # Inversion operator
coin_flip.h(0)
coin_flip.z(0)
 # Measure the qubit and save the outcome to the standard list
coin_flip.measure(0, 0)
 # implement the circuit on a local simulator
simulator = Aer.get_backend("Gasm_Simulator")
```

```
result = execute(coin_flip, simulator, shots=1024).result()
 # Print the results
counts = result.get_counts(coin_flip)
print(counts)
```

The measurement result is stored in a quantum circuit made up of one qubit and one classical bit created by this code. The circuit uses the Hadamard gate to superposition the qubit before using an oracle to flip it with a probability of 1/2. The qubit's amplitude in the initial state is then increased using an inversion operator. The circuit then measures the qubit and saves the outcome to the conventional bit. The algorithm increases the likelihood of measuring the target state (0 or 1) and decreases the probability of measuring any other state by repeatedly running this circuit. The outcome is determined by keeping track of how frequently each state is measured. The distribution that this code should provide for states 0 and 1 is roughly 50% for each, which is equivalent to the outcome of tossing a coin.

It is possible to quickly factor big composite numbers into their prime factors using Shor's algorithm, a quantum method. It can also be used to resolve the discrete logarithm issue in finite fields, which forms the foundation of numerous encryption techniques. The quantum version of the phase estimation technique is used to calculate the period of a function that translates a dice roll to the outcome of flipping a coin in order to apply Shor's algorithm to the problem of rolling dice. Quantum states are used to represent the probable outcomes of dice rolls and coin flips. For a total of 7 qubits, 6 qubits are utilized to represent the potential dice rolls and 1 qubit to represent the coin flip.

The function that converts a dice roll to a coin flip needs to be defined next. Using a controlled-NOT gate (CNOT), that flips the coin qubit if the dice roll is more than or equal to 4, can also accomplish this using a unitary operator. Once the input function has been defined, apply the quantum phase estimation method in order to assess the function's period. The input state is processed by a series of controlled unitary operations, and subsequently a quantum Fourier transform. Apply the information about the period of the function that is contained in the resulting state to find the outcome of the coin flip with a high degree of probability. When two particles are put into a single quantum state and prepared for quantum parallelism, they will always be in the same state when one particle is seen to be in a particular state. Here, the idea is demonstrated by tossing two coins. Flipping two coins would require two different calculations in traditional computing. However, quantum parallelism allows the quantum computer to calculate both coin flip results simultaneously, potentially saving time and processing power.

6.5.1 Future Applications and Challenges of Quantum Parallelism

Quantum parallelism includes an opportunity to change a wide range of industries, including machine learning, drug development, data encryption, and cryptography. In the domain of optimization, where quantum computers can be applied to work out demanding optimization problems that are challenging for classical computers, quantum parallelism has many interesting applications. Quantum parallelism, for instance, has been utilized to improve financial portfolios, airline timelines, and even the development of quantum algorithms themselves. Quantum parallelism is also used in the simulation of quantum systems, which is crucial for learning the actions of molecules, materials, and other quantum systems. Quantum parallelism may make it possible to replicate complex quantum systems more accurately and efficiently, which may lead to improvements in the fields of chemistry, materials science, and physics. Large-scale quantum systems are difficult for classical computers to model, however, quantum parallelism may be able to do this.

Before quantum parallelism can be fully understood, there are still important obstacles that must be overcome. Decoherence, which happens when a quantum system interacts with its surroundings and causes it to lose its quantum coherence and get entangled with it, is one of the main difficulties. The length of time that a quantum computation may be carried before errors become too substantial is constrained by decoherence, a key source of errors in quantum computation. Creating and maintaining large-scale entangled states that are necessary for quantum parallelism [7] is a further challenge. Maintaining the qubits' coherence and entanglement and performing operations on them without introducing errors becomes more challenging as the number of qubits grows. This necessitates the creation of novel approaches to quantum control, fault tolerance, and error correction.

Another difficulty is creating novel algorithms and applications that can benefit from quantum parallelism. It is not yet obvious which algorithms and applications will be the most beneficial or effective for quantum computers because the field is just beginning and is growing so quickly. This necessitates constant investigation and testing, as well as interaction between researchers from many disciplines. Despite these difficulties, many scientists feel that quantum computers will soon become a reality because of how helpful quantum parallelism could be in the future. We could potentially solve challenges that are currently unsolvable and acquire new knowledge about the nature of the universe by leveraging the strength of quantum parallelism.

6.6 QUANTUM COMPUTING SYSTEMS

Computers that perform computations using quantum mechanics are known as quantum computing systems. These systems are designed to address problems which are too difficult for conventional computers to deal with [1]. Quantum computers use entanglement and superposition, two examples of quantum mechanical phenomena, to carry out calculations. Although they are still in the early phases of research, quantum computers have already showed promise in solving some issues that traditional computers find challenging or impossible. D-Wave platforms, IBM quantum Systems, and the recently unveiled IBM Quantum System One are just a few examples of quantum computing platforms [5]. One of the most well-known and potent quantum computing platforms in the world, IBM quantum systems allow programmers to develop novel solutions to issues.

Quantum computers improve the performance of certain types of complex calculations by using special features of quantum mechanics like superposition and entanglement (Table 6.1).

TABLE 6.1
Two Absolutely Distinct Mechanisms for Processing Information: Classical Computing and Quantum Computing [10]

Aspect	Classical Computing	Quantum Computing
Key unit	Bits => (0 or 1)	Qubits => (superposition of 0 and 1)
Data processing	Binary arithmetic and logic operations	Quantum gates and entanglement
Computational model	Deterministic model (the parameter and starting values completely define the model's output)	Probabilistic model (randomness is incorporated into the entire procedure because of superposition and entanglement)
Time complexity	Time taken 2^T (on average, polynomial/exponential)	Time taken T (possibly exponential velocity for particular issues)
Speed	Minimal speed for some operations that can be parallelized	Exponential speed of some quantum techniques
Problem scope	Useful for a variety of common problems	Designed to address particular issues
Error sensitivity	Fault-resistant; uses techniques for error correction	Subject to inaccuracies; needs to be rectified
Scalability	Scalable to more difficult issues and resource-intensive	Qubit coherence introduces difficulties with scalability.
Hardware requirements	Classical transistors processor architectures	Qubits and specialized quantum processors [7]
Maturity	Well-established technology	Early stages of development
Power consumption	Normally more power-efficient	Quantum computing often needs particularly low temperatures

While entanglement is a process in which two or more particles can become entangled, meaning their quantum states become associated in a way that goes beyond conventional correlations, superposition discusses to the capacity of quantum objects to occur in more than one state concurrently (Figure 6.5).

Qubits, where represent the quantum model of the conventional bits utilized by classical computing, are employed in quantum computing systems to take advantage of these qualities. Multiple qubits can become entangled and exist in a superposition of both 0 and 1 at the same time, allowing for the simultaneous processing of enormous amounts of data. There are still major challenges in developing and scaling quantum computers for usage, even if these characteristics make quantum computing systems more effective than classical ones for some tasks, such as factorization challenges that are challenging for classical computers.

6.6.1 Qubit

A qubit, which is equivalent to the traditional binary bit in classical computing, is the basic quantum information unit. It can represent a zero, a one, or any superposition of these two states. It is the quantum equivalent of a bit. Superposition and entanglement allow qubits to occur in many states at once, unlike classical bits, which can

FIGURE 6.5 Quantum computing.

only ever be in a 0 or 1 state. The spin of an electron, which has two levels that can be interpreted as spin up and spin down, is an instance of a qubit (Table 6.2).

In conclusion, the fundamental building blocks of quantum information are single qubits, which can exist in superposition. However, many qubits are a more potent computational resource because they can exist in complicated entangled states and take advantage of quantum parallelism. Utilizing the exponential increase in computational power with the number of qubits, multi-qubit systems and quantum algorithms are frequently used in quantum computing to solve complicated problems.

TABLE 6.2
Shows the Difference between Single Qubit and Multi-Qubit

Aspect	Single Qubit	Multiple Qubits
Key unit	One quantum bit (qubit)	Two or more qubits (entangled or not)
State representation	Presented by Bloch sphere	Presented by multi-qubit state vector
Quantum gates	Hadamard, Pauli-X, Pauli-Y, Pauli-Z, etc.	Controlled-NOT, SWAP, Toffoli, etc.
Entanglement	Not applicable	Vital to quantum computing
Quantum parallelism	Single-qubit operations have a constraint	Permits exponential speed in some situations
Measurement	Gives a typical bit outcome of 0 or 1	Gives multi-qubit state outcome
Quantum error correction	Not directly involved	Required to keep qubit coherence
Quantum algorithms	Confined to specific algorithms	Capacity to execute efficient algorithms
Communication and complexity	Usage is minimized in communications tasks	Possibility of solving challenging issues
Quantum supremacy	Not applicable	potentially permits quantum supremacy
Scalability	Simple to deal with	More qubits lead to higher levels of complexity

6.6.2 Superposition

Superposition is a key idea in quantum mechanics and describes how a quantum system can exist in many states or locations simultaneously up until it is measured. Contrasted with classical mechanics, where concepts like location and momentum are always clearly defined, in quantum computing, a qubit is any quantum system that is capable of existing simultaneously in the quantum states of 0 and 1. Superposition is a crucial aspect of quantum computing that enables massive parallel data processing. In other words, quantum computers can execute some operations ten times faster than classical computers by making use of superposition. Superposition enables complicated calculations that would be impossible for conventional computers by allowing quantum things to simultaneously exist in multiple states or positions.

Quantum superposition is the ability of a quantum system to exist in several states successively up until it is measured. All of these scenarios combined, with a complex integer defining the amount in each configuration, form the most general scenario. But in addition to heads and tails, a quantum coin also has an additional feature called quantum phase. The coin can still be heads or tails in the quantum phase, but it can also be in a superposition of both heads and tails. A classical coin toss does not involve quantum phase or superposition. A traditional coin toss starts with either heads or tails up, flips a number of times depending on who or what is performing it, and then lands with either heads or tails up. It is, thus, conceivable to predict a classical decision with arbitrary precision by carefully measuring all of those classical actions (Figures 6.6 and 6.7).

However, any significant number of classical actions will also always involve some quantum actions. This is because the tiny universe is not classical but rather quantum and unpredictable. Until the macroscopic coin lands as heads or tails, a

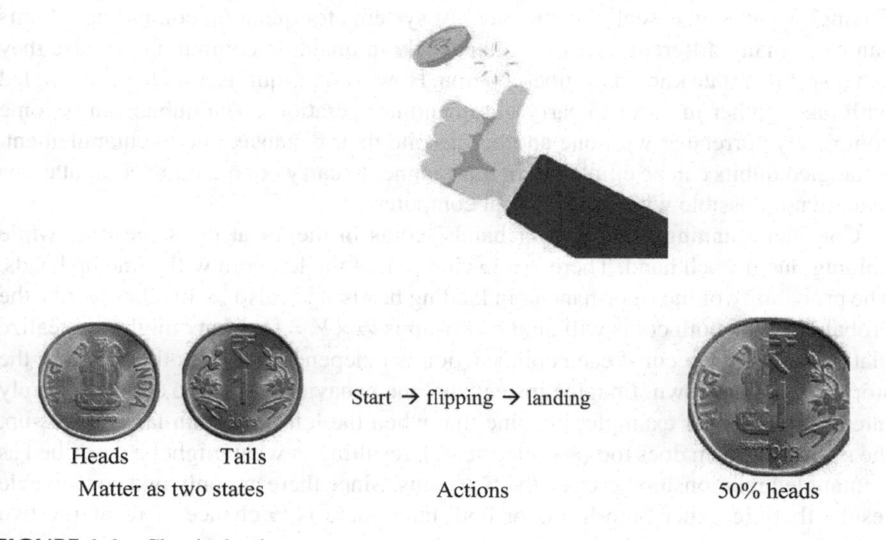

FIGURE 6.6 Classical coin toss.

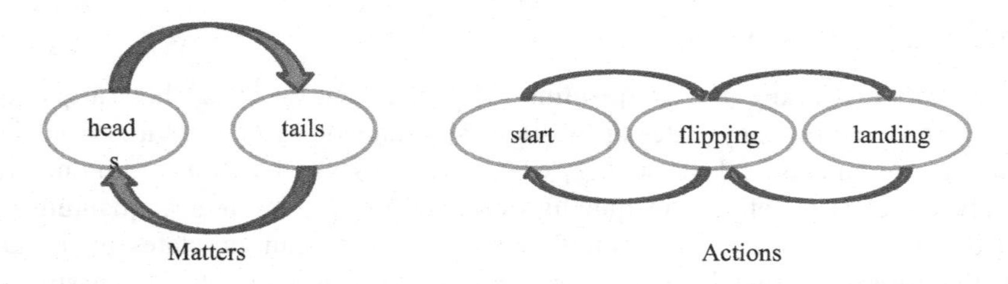

FIGURE 6.7 Quantum coin toss.

dephased quantum coin will remain within a microscopic superposition. There are no measurements that can accurately forecast the result of a quantum coin flip from a superposition state better than 50%, in contrast to the precise classical measurements that can do so. In our quantum real world, the conventional coin toss actually does not exist. Instead, many conventional macroscopic activities, such as tossing a coin, will eventually access several microscopic quantum consequences of quantum phase noise. The underlying microscopic quantum phase noise of physical reality serves as a useful macroscopic template for the classical random noise of chaos.

Keep in mind that, in addition to the quantum coin, superpositions of the quantum acts of starting, flipping, and landing also exist. Quantum phase noise will also be entangled by several operations, such as flipping or neural impulses, and as a result will be subject to the constraints of quantum knowledge's level of uncertainty. It is difficult to understand the ideas of free will and free choice because it can never accurately know something that precedes another or forecast a feeling that comes after another.

6.6.3 Entanglement

Entanglement is an essential component of systems for quantum computing. Qubits can be in many different states simultaneously in quantum computing because they can reside in a state known as superposition. However, the qubits need to be entangled with one another in order to carry out quantum operations. The qubits can become completely correlated with one another despite their distance, due to entanglement. Entangled qubits can be employed in this manner to carry out quantum computations that are not possible with conventional computers.

Consider spinning both of your hands' coins in the air at the same time while holding one in each hand. There is a ½ chance that the left coin will come up heads. The probability of the right-hand coin landing heads-up is also ½. In other words, the probability that both coins will land heads-up is $\frac{1}{2} \times \frac{1}{2} = \frac{1}{4}$. Many might not realize that this can only occur if each coin's action is independent of the other because the story is so well-known. Imagine instead that the behavior of the two coins is entirely interdependent. For example, imagine that when the left-hand coin lands heads-up, the right-hand coin does too (and vice versa), resulting in what might be described as a entangled relationship between the two coins. Since there are only two conceivable results that are either both heads or both tails, there is ½ chance of receiving two heads in that scenario (Figures 6.8 and 6.9).

FIGURE 6.8 Independent coins.

FIGURE 6.9 Entangled coins.

The below sample code, shows two qubits and two conventional bits are first used to build a quantum circuit.

```
# Import necessary libraries
from qiskit import QuantumCircuit, a, exe
from qiskit.visualization import plot_histogram
 # Build a quantum circuit having two Calssical_bits and two
   qubits
dice_coin_circuit = QuantumCircuit(2, 2)
 # Put them in superposition by apply a Hadamard gate to both
   qubits
dice_coin_circuit.h(0)
dice_coin_circuit.h(1)
 # Applying a CNOT gate and form entanglement between the qubits
dice_coin_circuit.cx(0, 1)
 # Qubit is evaluated and store the results in a conventional
   bit
dice_coin_circuit.measure([0, 1], [0, 1])
 # Execute the circuit by used QASM simulator at the backend
   to display the result
simulator = a.get_backend("qasm_simulator")
result = exe(dice_coin_circuit, simulator, shots=1024).result()
 # Histogram of results
plot_histogram(result.get_counts(dice_coin_circuit))
```

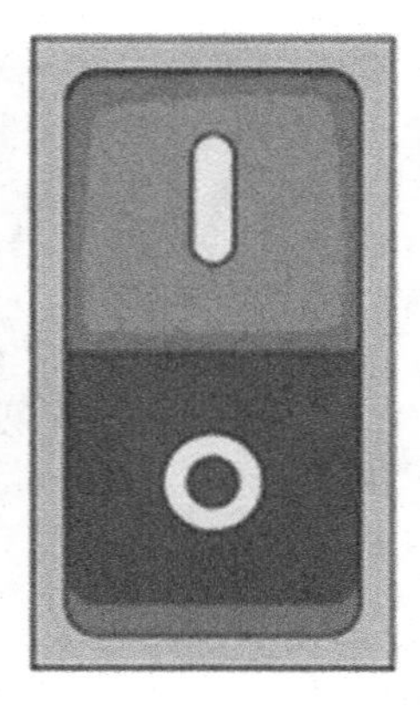

FIGURE 6.10 Classical bits.

Then, after both qubits are placed in superposition using Hadamard gates, CNOT gates are used to establish entanglement between the two qubits. After measuring the qubits, all the data are then recorded on the corresponding classical bits. Finally, a quantum assembly language simulator (QASM) backend is used to execute the circuit and obtain the results. The findings display the likelihood of each potential occurrence, simulating the outcome of tossing a coin and rolling a die in superposition.

Bits that can only exist in one of two states, 0 or 1, are the basic building blocks of traditional information systems. Qubits, two-level quantum-mechanical devices that can simultaneously exist in a superposition of 0 and 1, are the core components of quantum information systems, and are used to store information. Quantum systems can carry out some computations ten times more quickly than classical systems due to this superposition feature (Figures 6.10 and 6.11).

Entanglement is a feature of quantum information systems that allows two or more qubits to be connected in a way that makes one qubit's state depended on the states of the others. Multiple possible applications for this aspect include quantum computing and quantum communication. However, quantum information systems are also stronger than classical ones and are exposed to interference and noise from the outside world. A quantum system's state can also be changed by measurement in a way that is not feasible in classical systems. When developing and managing quantum information systems, such factors must be taken into considerations. Calculations can be executed simultaneously on quantum computers (Figures 6.12 and 6.13).

Quantum information systems can be made more secure by applying quantum cryptography, which encodes data using the laws of quantum physics in an approach that is likely to be hard to decode. Despite the fact that classical and quantum information systems can both store and transport information, quantum information systems have special qualities that could offer them an important benefit over classical systems in specific situations (Figure 6.14).

The number of trials or simulations required will determine how long it takes to replicate the rolling of a dice or flipping of a coin using classical computing. The likelihood to obtain heads or tails on a single coin flip is 0.5. This can be replicated

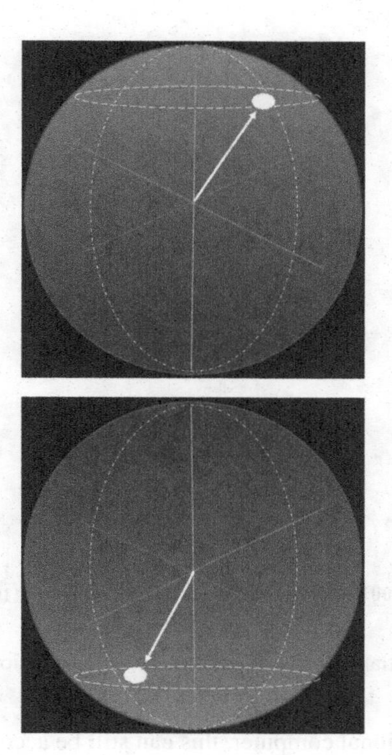

FIGURE 6.11 Qubit.

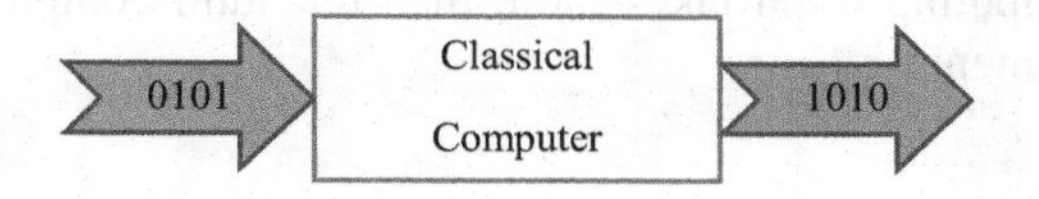

FIGURE 6.12 Classical Boolean logic.

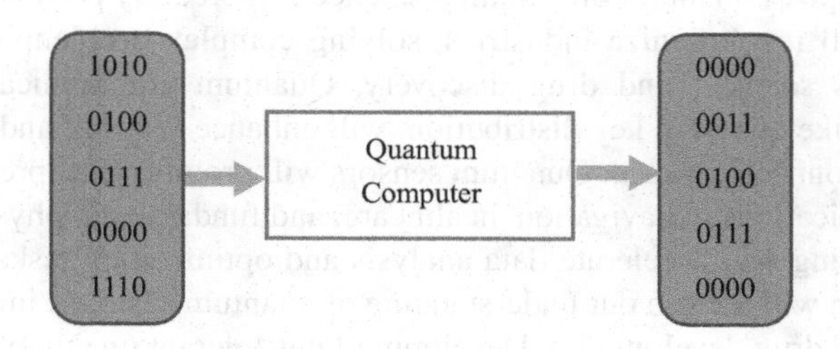

FIGURE 6.13 Quantum logic.

on a traditional computer using a Boolean flip, and it would take very little time (nanoseconds or less) to complete. The computing cost of simulating numerous dice rolls or coin tosses can increase. For instance, if you wanted to imitate rolling two dice 1000 times, you would have to simulate rolling two dice 1000 times and add up

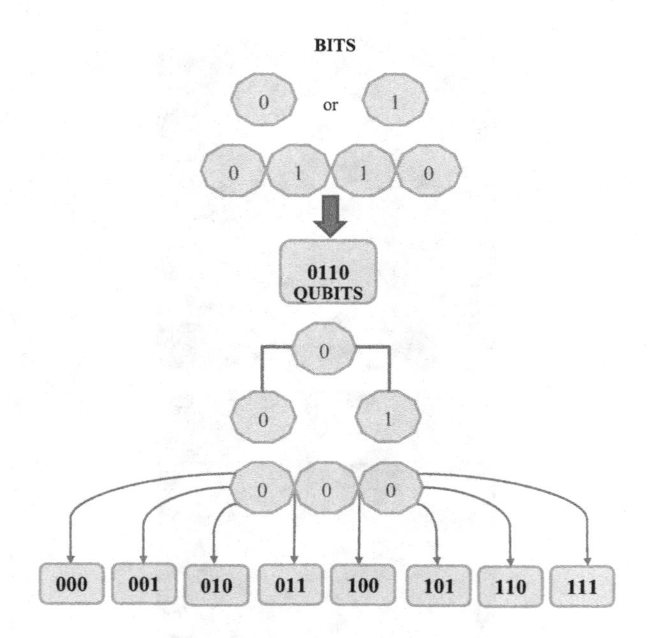

FIGURE 6.14 Exponential capacity of qubits to store information.

the faces. On a conventional computer, this can still be accomplished rather quickly and would only require a few milliseconds or less, depending on the quantity of simulations performed. A study that measured the quantity of trials executed found that classical computing often takes more than quantum computing to perform a probability experiment.

6.7 SUMMARY AND FUTURE SCOPE

The future scope of quantum information science is incredibly promising. Quantum computing will revolutionize industries, solving complex problems in cryptography, materials science, and drug discovery. Quantum communication, through technologies like quantum key distribution, will enhance security and privacy in an increasingly connected world. Quantum sensors will provide ultra-precise measurements for applications in navigation, healthcare, and fundamental physics. Quantum machine learning will accelerate data analysis and optimization tasks, while quantum simulation will deepen our understanding of quantum systems, impacting material design and drug development. Developing fault-tolerant quantum hardware and algorithms will make quantum computing more practical. Quantum education and policy initiatives will address the growing demand for skilled professionals and ethical considerations. Overall, quantum information science's future will see transformative advancements across various domains, reshaping technology, research, and daily life.

REFERENCES

1. Manjunath, T.N. et al. 2022, "A Survey on Machine Learning Techniques Using Quantum Computing", *Fourth International Conference on Emerging Research in Electronics, Computer Science and Technology (ICERECT)*, https://doi.org/10.1109/ICERECT56837.2022.10059764
2. Pattanayak, Santanu 2021, *Quantum Machine Learning with Python*, Springer Science and Business Media LLC, https://doi.org/10.1007/978-1-4842-6522-2
3. Kasirajan, Venkateswaran 2021, *Fundamentals of Quantum Computing*, Springer Science and Business Media LLC, https://doi.org/10.1007/978-3-030-63689-0
4. Heusler, Stefan et al. 2022, "Aspects of Entropy in Classical and in Quantum Physics", *Journal of Physics A: Mathematical and Theoretical*, Volume 55, Number 40, https://doi.org/10.1088/1751-8121/ac8f74
5. Córcoles, Antonio D. et al. 2020, "Challenges and Opportunities of Near-Term Quantum Computing Systems", *IEEE Explore*, Volume 108, Issue 8, https://doi.org/10.1109/JPROC.2019.2954005
6. Hofmann, Holger F. 2005, "Quantum Parallelism of the Controlled-NOT Operation: An Experimental Criterion for the Evaluation of Device Performance", *Physical Review A*, Volume 101, Issue 12, https://doi.org/10.1103/PhysRevLett.101.12050
7. Chi, Y. et al. 2022 "A Programmable Qudit-Based Quantum Processor", *Nature Communications*, Volume 13, Issue 1166, https://doi.org/10.1038/s41467-022-28767-x
8. Schade, Robert et al. 2022, "Parallel Quantum Chemistry on Noisy Intermediate-Scale Quantum Computers", *Physical Review Research*, Volume 4, Issue 3, https://doi.org/10.1103/PhysRevResearch.4.033160
9. Zhou, X. 2023, "Dynamical Behavior of Quantum Correlation Entropy Under the Noisy Quantum Channel for Multiqubit Systems", *International Journal of Theoretical Physics*, Volume 62, Issue 23, https://doi.org/10.1007/s10773-022-05270-z
10. Pizzi, Andrea 2022, "Bridging the Gap between Classical and Quantum Many-Body Information Dynamics", *Physical Review B*, Volume 106, Issue 21, https://doi.org/10.1103/PhysRevB.106.214303
11. Yuan H.N. et al. 2022, "Quantum Magnonics: When Magnon Spintronics Meets Quantum Information Science", *Physics Reports*, Volume 965, https://doi.org/10.1016/j.physrep.2022.03.002

7 Quantum Machine Learning Approaches

Mangalraj Poobala and Ganesh Kumar Natarajan
GITAM University, India

Iniyan Shanmugam
SRM Institute of Science and Technology, India

Justin Vargese
GITAM University, India

7.1 QUANTUM COMPLEXITY

Quantum complexity is a fundamental concept to identify the complexity of any problem in correspondence to quantum computing [1]. Several parametric resources, such as time, space, and operations, are considered to evaluate the complexity of the problem. The quantum mechanism allows machines to run in multiple states simultaneously. This simultaneous process requires quantum superimposition on the target machine for effective execution [2]. The significant difference in evaluating the complexity of classical and quantum computing is identifying the parallelism and relationship between the execution threads.

The quantum superposition property allows complex calculations to be more efficient, and several quantum algorithms have been implemented to perform such calculations [3]. Quantum algorithms help to solve complex problems in a fast-paced manner compared to classical algorithms. Entanglement is another exciting concept used in quantum computing that unveils correlations in integral parts [4]. Quantum complexity for any complex problems such as P and NP will be effectively calculated by bounded error quantum polynomial time (BQP). The BQP calculates bounded error probability in correspondence to quantum operations [5]. Quantum complexity allows developers to understand the pros and cons of quantum computation.

7.2 FEATURE MAPS

Mapping of actual data from a global space to a feature space with the help of feature maps improves or solves any non-linear problems. Feature maps seek to map non-separable input data to separable features for better training purposes, as shown in Figure 7.1

DOI: 10.1201/9781003429654-10

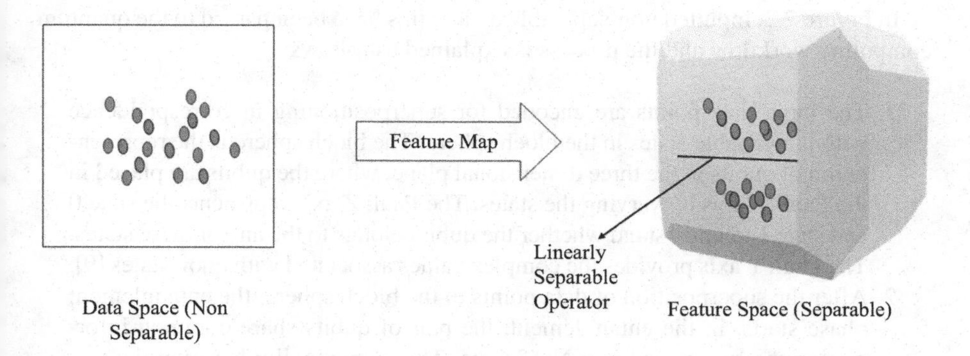

FIGURE 7.1 Conventional feature map utilization.

Conventionally feature maps are mathematical functions which help to map the input data to the n-dimensional feature space where the features can be linearly separable [6]. The separator could be a linear one, or a non-linear one, but the feature points are separable [7]. The encoding operation generates the states of the input and will be denoted through a Bloch sphere through the quantum superposition property. As mentioned earlier, the entanglement property helps to identity the correlations between the data points and the enhanced feature map will be generated as shown in Figure 7.2 Finally, quantum hardware is used to process mapped feature points as an input to the classical machine for training a machine learning algorithm [8]. The schematic diagram in Figure 7.2 depicts the overall utilization quantum feature map in real time.

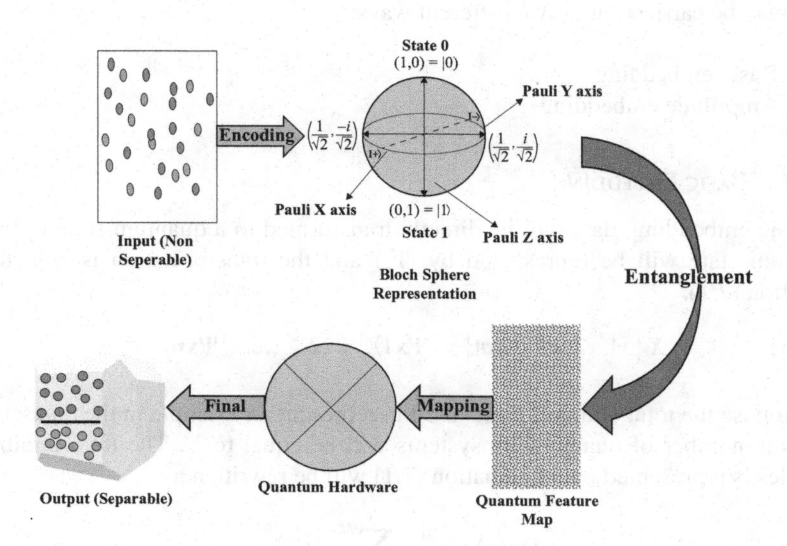

FIGURE 7.2 Workflow schematic diagram of utilization of quantum feature map.

In Figure 7.2, inputted non-separable data points have been parsed to the quantum computing workflow and the process is explained as follows:

1. The input data points are encoded for superpositioning in correspondence with the variable states in the bloch sphere. The bloch sphere is the representation of qubits in the three dimensional plane, where the qubits are placed in the Pauli X axis by varying the states. The Pauli Z axis represents the state 0 and state 1 to understand whether the qubit belongs to the any of these states. The Pauli Y axis provides the complex values associated with qubit states [9].
2. After the superposition of data points to the bloch sphere, the entanglement phase starts. In the entanglement, the pair of qubits share common information through a common NOT Gate. The commonality between the two qubits generates the new feature vector for the input data [10].
3. Once the feature vector is generated for the input, the feature map for the quantum computing will be generated. The quantum feature map is the combination of feature vectors and quantum states.
4. Feature maps are encoded and processed through quantum hardware for faster processing and precision results.

Quantum complexity always relies on the entanglement and quantum hardware processing; however, many complex problems can be effectively handled using quantum computing.

7.3 QUANTUM EMBEDDING

Quantum embedding is defined as the process of representing classical data points or data to a quantum state by parsing the input through a feature map [11]. The embedding may be carried out in two different ways:

1. Basic embedding
2. Amplitude embedding

7.3.1 Basic Embedding

In basic embedding, data will be directly transformed to a quantum state [12]. The quantum state will be represented by $|\Psi_x\rangle$ and the transformation is depicted in Equation (7.1).

$$X\{x1, x2, x3\ldots, xn\} = |\mathbf{\Psi x1}\rangle, |\mathbf{\Psi x2}\rangle, \ldots\ldots, |\mathbf{\Psi xn}\rangle \tag{7.1}$$

Suppose the total number of bits used to represent an example in the dataset is N, then the number of quantum subsystems will be equal to N. The total number of samples is represented by M. Equation (7.1) will be rewritten as

$$|\mathbf{Data}\rangle = \frac{1}{\sqrt{M}} \sum_{m=1}^{M} |x^m\rangle \tag{7.2}$$

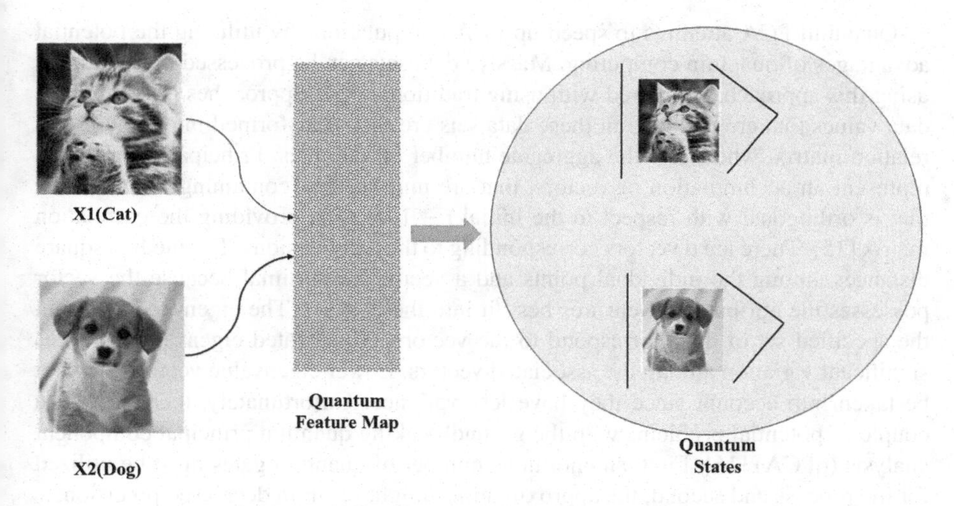

FIGURE 7.3 Basic embedding of classical data to quantum states.

Figure 7.3 provides better understanding of quantum embedding in an uncomplicated way.

7.3.2 Amplitude Embedding

The amplitude of each and every state is embedded to generate the quantum states in the Hilbert Quantum Space. The amplitude-based embedding is depicted through Equation (7.3).

$$|\mathbf{Data}\rangle = \sum_{m=1}^{2M} a_m |m\rangle \tag{7.3}$$

where, a_m is the normalized concatenated data, and where m is the computational basis state.

There are similar state of the art embeddings available such as angle encoding, rotation encoding, Qsample encoding, and divide, Conquer encoding, and so on [13].

7.4 QUANTUM PRINCIPAL COMPONENT ANALYSIS (qPCA)

A quantum approach referred to as quantum principal component analysis (qPCA) harnesses quantum computing to reduce the number of dimensions in a dataset and determine the features that are most important. For analyzing information and compression in conventional machine learning applications, PCA remains a widely employed approach. During conventional PCA, the approach identifies a collection of principal components that is, orbifold directions resulting in the data varying most substantially. These elements enable reduction of dimensions and the extraction of features by capturing each dataset's vital information.

Quantum PCA attempts to speed up PCA computations by utilizing the potential advantages of quantum computing. Massive data sets can be processed more quickly using this approach, compared with using traditional PCA approaches [14]. Variable data values that are included in these data sets are then transformed into an X n correlation matrix, where n is the aggregate number of variables. Principal components represent an accumulation of vectors, that are unit vectors, containing an *i*th vector that is orthogonal with respect to the initial $i - 1$ vectors, providing the correlation matrix [15]. There are d vectors corresponding to the d dimensions. The median square distances among the individual points and a vector are minimal because the vector possesses the attribute that ensures best fit into the data set. The eigenvectors within the specified set of data correspond to the vectors. An elevated eigenvalue signifies significant variation among the associated vectors. Lower eigenvalue vectors may not be taken into account since they have less variance. Unfortunately, there remain a couple of potential problems with the groundbreaking quantum principal component analysis (qPCA) [16]. First, an enormous number of quantum gates must be utilized for the process and second, the approximations might result in decreased precision.

The algorithm has been examined by multiple researchers and found to be a precise algorithm, which means that when the technique is used, it yields precise outcomes rather than estimated outcomes with low complexity. This attribute renders it a more effective algorithm than the technique explained in the work of some scholars in current research based on quantum singular value threshold (qSVT). Only these two approaches are independent of controlled operations as well as initialization matrices [17]. The operational stages of the low complexity qPCA algorithm are:

1. Phase estimation: In this stage, the eigenvalues are abstracted to the quantum register.
2. Unitary operation: In this stage, the trivial eigenvalues are separated out from the quantum register.
3. Unitary controlled operation: In this stage, unitary controlled operation and an auxiliary qubit are used to determine the existence of the observed quantum bits' eigenvalues belonging to a primary component.
4. Unitary reverse operation: In this stage, value-storing registers deemed superfluous are removed.
5. Measurement: This stage quantifies the qubits.
6. The second phase estimation: In this stage, the quantum state can be attained.

7.5 QUANTUM LINEAR MODELS IN MACHINE LEARNING

Linear regression models have been widely used to predict the targets in correspondence with the predictors. The linear correlation of x and y variables may be identified as shown in Equation (7.4).

$$y = ax + b \tag{7.4}$$

where a and b represent the slope and intercepts of the linear model derived from the correlation of x and y variables [18]. However, determining the optimal value

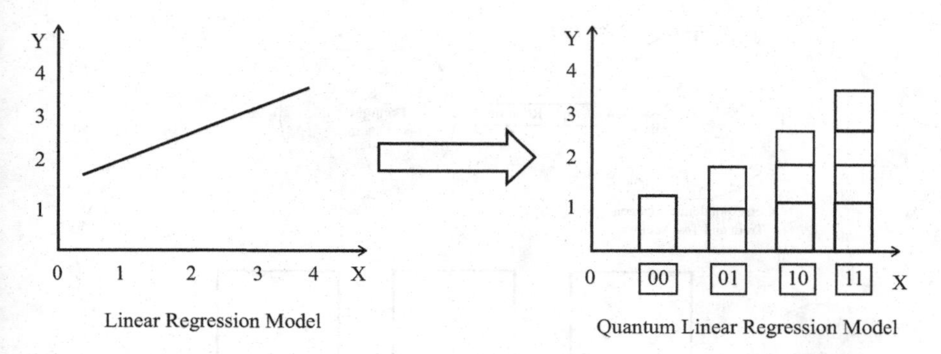

FIGURE 7.4 Quantum regression model.

for a and b is always tedious when the data size increases. Quantum linear models are exceptional at addressing the issue of identifying the best optimum in a more extensive data set. For better understanding, Figure 7.4 depicts the linear model vs. the quantum linear model.

Conceptually, quantum linear regression incorporates the states concerning every qubit, and is plotted as a discrete map in Figure 7.4 So, mathematically, x and y have been converted into qubits and can be normalized as shown in Equation (7.5):

$$|y\rangle = \frac{1}{C_y} y \tag{7.5}$$

In Equation (7.5), $\frac{1}{C_y}$ is considered as the normalized factor. The mean square error is reduced to obtain more precision in the final model [19]. The cost function reduces the mean square error. However, different researchers have proposed many methods to estimate cost function for a quantum linear regression [20]. Polynomials achieve the extended quantum regression technique based on higher order fit.

7.6 QUANTUM CLUSTERING

Reference [8] clustering is recognized as an unsupervised learning algorithm and is mainly useful in identifying the underlying pattern in an input data stream. The patterns are primarily utilized to determine the behavior of the data stream in terms of mathematical notions and statistical inferences, and, in some instances, it is used as a dimensionality reduction technique for hybrid modeling [21]. The underlying principle for clustering is to determine the similarities between data or examples. However, when the data stream is extensive, the redundant computation on classical machines fails to achieve the best results [22]. Quantum computing resolves these issues through its hardware implementation of clustering algorithms for large-scale data streams. Conventional K-Means and K-Median clustering is utilized next to explain and discuss the implementation of quantum clustering and the effectiveness of the algorithms.

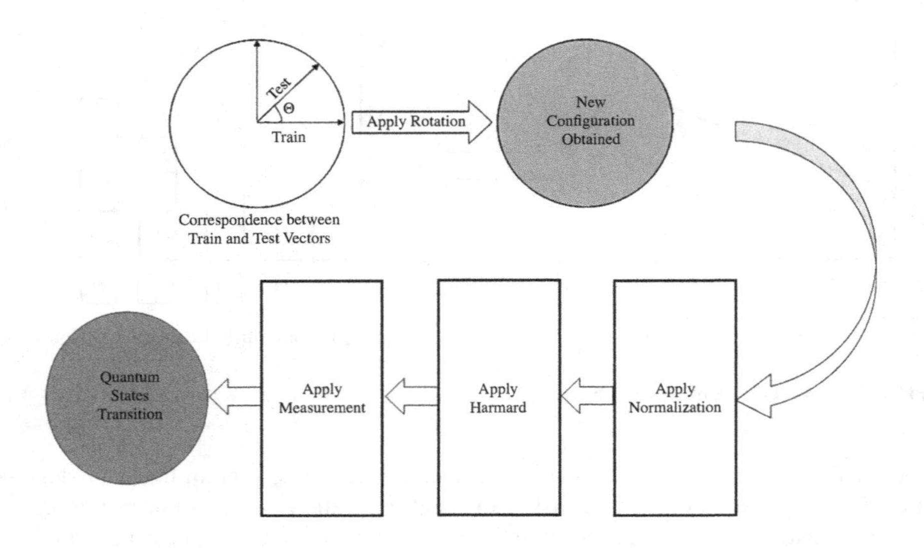

FIGURE 7.5 Quantum clustering process.

7.6.1 Quantum K-Means Clustering

Quantum K-Means clustering is performed on training and test data and the correspondence between each vector is identified to obtain the best accuracy [23]. Alternatively, in classical K-Means clustering, the training input data is considered for determining test accuracy. The flow diagram in Figure 7.5 depicts the process of K-Means clustering in a detailed way.

After quantum embedding is performed on the train and test vectors, the correspondence between the train and test vectors is established, with respect to the degree of position [24]. A rotation operation is performed to obtain a new vector from the original and to obtain better correspondence between the input vectors. After obtaining the newly configured Qubits, the normalization is performed on the vectors [25]. The Hadamard operation is performed on the normalized vector on the most significant qubits. Measurement operations are performed to identify the probabilities of quantum states. The whole process is repeated to identify the relative difference of the quantum states, eventually leading to accurate clustering results by inferencing the variational states.

7.6.2 Quantum K-Median Clustering

K-Median is a conventional clustering technique where K denotes the number of clusters required from the input data. Once the execution starts, K-points are selected, and the proximity of the data points to the centroids is calculated and forms a cluster. This process will repeat recursively until it reaches the stable centroids. The recursive nature of the process is time consuming, and the complexity increases exponentially when the data size increases. To address this effectively with large datasets, quantum K-Median implementation helps for faster convergence and to decrease the complexity [26]. Multiple versions of K-Median clustering have been introduced, and the most famous is the Diffie-Hellman algorithm (DH) algorithm, developed by Aïmeur et al. [27], which is widely in used to mitigate the minimization problem in K-Median clustering.

7.6.3 Quantum Hierarchical Clustering

Quantum hierarchical clustering is one of the leading methods used for clustering and identifying the most prominent features of data. Hierarchical clustering recursively identifies correlations among the training data through proximities at every level [28]. The level denotes the prominent feature of the data which helps to identify the clusters. Proximities are calculated using distance-based methods. Hierarchical clustering is depicted in Figure 7.6.

Hierarchical clustering is carried out in two different ways. One way is the agglomerative method (bottom-up approach) and the other is the divisive method (top-down approach). The Euclidian distance between the data points is considered for clustering purposes. In the divisive method, the data set is classified into multiple clusters, while in the agglomerative method, the multiple clusters form into a single cluster. The levels represent the important attributes possessed by the dataset. In Figure 7.6, we notice that the data points (11, 8, 1, and 2) form cluster 1, (10, 7, and 3) form cluster 2, and (9, 6, 4, and 5) form cluster 3. Every level represents the attribute values and is based on the values of the clusters that will be formed. However, when the dataset size is increased, finding the optimal distance measure between the data points is a tedious process. To identify the optimal distance between the data points for better convergence, quantum computing is employed [29]. The implementation of hierarchical clustering using quantum hardware is discussed

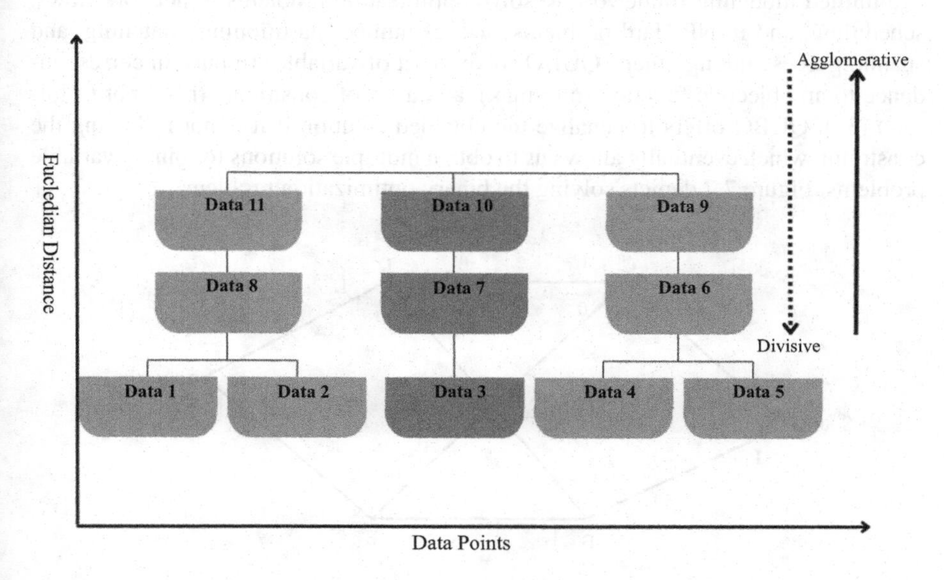

FIGURE 7.6 Hierarchical clustering.

below. In quantum hierarchical clustering the same concept of quantum clustering process is applied, as shown in Figure 7.5.

7.7 THE HARROW–HASSIDIM–LLOYD (HHL) ALGORITHM

Solving any linear system is very important in the field of engineering. Essentially, linear systems yield no solution, multiple solutions, or a single solution. No-solution linear systems are called inconsistent, multiple-solution linear systems are called consistent and dependent, because the solutions are dependent on other solutions, and the single-solution linear system is called consistent and independent, because the solution is independent of any inputs or influences [30, 31]. All linear systems follow the homogeneous principle as shown in Equation (7.6).

$$\left[ay1(t)+by2(t)\right]=T\left[ax1(t)+bx2(t)\right]=aT\left[x1(t)\right]+bT\left[x2(t)\right] \tag{7.6}$$

Since solving a linear system has severe complexity issues in classical computing, quantum computers are much more successful in solving linear systems to obtain a consistent and independent output. The scientists Hassidim, Harrow, and Lloyd developed an algorithm to solve the linear systems effectively, known as the HHL algorithm.

7.8 QUADRATIC UNCONSTRAINED BINARY OPTIMIZATION (QUBO)

The quadratic unconstrained binary optimization (QUBO) method is widely used to solve combinational optimization problems in various fields. The QUBO method is a unified modeling framework to solve optimization problems in network flows, scheduling, and in NP-Hard problems such as number partitioning, matching, and spanning trees, among others. QUBO solves a set of variables (input), in correspondence to an objective function (min/max), and a set of constraints (need not to follow) [32]. QUBO offers to penalize the obtained solution if it is not following the constraint, which eventually allows us to obtain multiple solutions for binary variable problems. Figure 7.7 depicts solving the binary optimization problem.

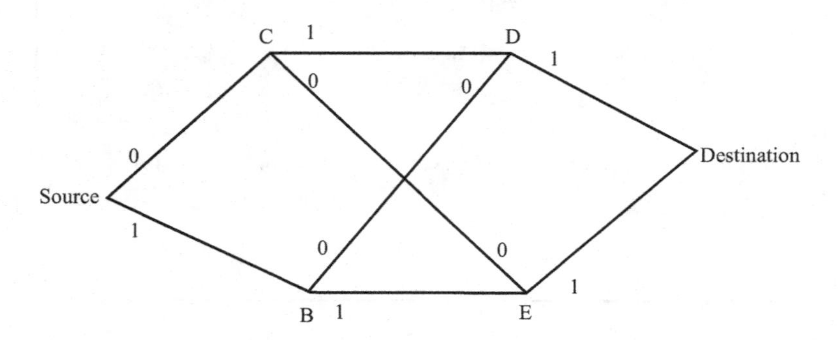

FIGURE 7.7 Binary optimization problem (path finding).

The binary optimization problem for path identification under constraints yields the following result [33]. Constraint Binary optimization always strongly relies on the constraint being satisfied and yields the best-optimized result as shown below.

$$\text{Path} = \{\text{Start}, \text{B}, \text{E}, \text{Destination}\}$$

However, there are cases when the constraints need not be satisfied when identifying the optimal result. Such problems are called unconstrained binary optimization. For the problem mentioned in Figure 7.7, we get possible solutions as follows:

$$\text{Path} = \{\text{Start}, \text{B}, \text{D}, \text{Destination}\} \text{ and } \{\text{Start}, \text{C}, \text{E}, \text{Destination}\} \text{ and so on.}$$

We obtain multiple solutions by violating the constraints at certain states, in such cases QUBO will penalize the solutions. The penalization will be carried out by generating a Q matrix, where the Q matrix is a symmetric one [34]. After penalizing, whichever solution received the least penalty, and is the most feasible solution, is considered as the optimal solution.

Further, QUBO is used to solve NP-complete problems by identifying the optimality in short polynomial time. The QUBO is mathematically defined as shown in Equation (7.7):

$$\text{min/max } y = x^t Q x \tag{7.7}$$

where x is the predictor variables, and Q is the square matrix of constants.

Finally, QUBO allows quantum computers to find solutions which satisfy those constraints, without forcing the constraints to hold true. This adds significant value in problems where multiple results offer deeper insights and more discernment for making better business decisions.

7.9 QUANTUM SUPPORT VECTOR MACHINES

A support vector machine (SVM) is a traditional classification algorithm. The support vector machine separates inseparable data points into linearly separable classes with the help of a hyperplane. The hyperplane is considered the hypothesis function, which linearly separates the classes in the feature space [35]. The hyperplane can be generated using hard and soft margin techniques. Soft margin techniques are widely employed since they offer dynamic solutions to find optimal solutions. In soft margin techniques, the error variable, called the slack variable, is introduced. This violates the constraints and allows certain misclassifications to arrive at the best solution, which is similar to QUBO as discussed in Section 7.8. The best solution is identified by discarding errors and increasing margins. The dual problem is one of the major issues that needs to be addressed in SVM. In the dual problem, the Lagrange multipliers are used to maximize the objective function (to maximize the margins). To achieve this, a kernel trick is used to compute the similarity between the samples. The kernels could be linear, polynomial, Gaussian RBF, or Sigmoid, and the type of

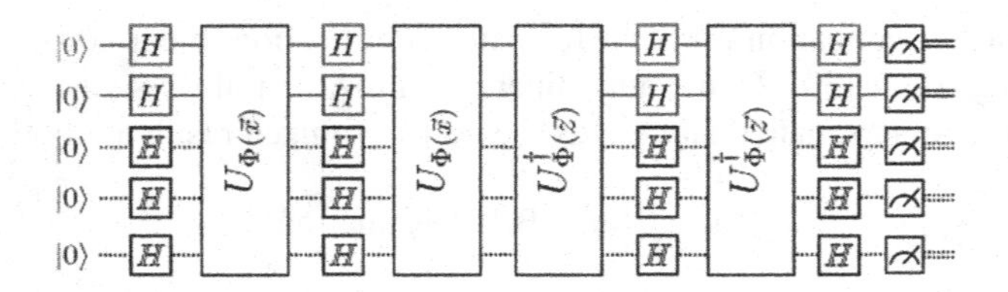

FIGURE 7.8 Circuit generated for extracting information from a quantum kernel.

kernel is chosen based on the feature space distribution [36]. Then, to identify the similarity of the samples, the similarity is calculated in quantum machines as shown in quantum clustering. If we have 40 samples, the kernel will be 40 × 40 in the quantum machine implementation. Once the quantum kernel is generated, the information from the kernel needs to be extracted and parsed to the classical system for final classification. To extract information from the quantum kernel, the circuit shown in Figure 7.8 is now utilized in this cutting-edge method.

The circuit shown in Figure 7.8 is utilized to identify the overlap of two states. By estimating the overlap, the frequency of 0^n is calculated because the initial state is 0. To generate these quantum circuits for the kernel, some researchers have generated automatic-circuit-generating schema which are currently available art [37]. Experimentally, the classical and quantum SVM does not have diversified accuracies, but the runtime complexity is decreased, and several researchers have proved this through analysis.

7.10 SOLVING NP-HARD PROBLEMS

The NP-Hard problems are solved in non-deterministic polynomial time which yields decisions. The NP-Complete problems also yield decisions but with polynomial time in a non-deterministic machine, reflecting the efficiency of identification of solutions. Routing is one of the best examples for a NP-Complete problem, where multiple solutions are available, and we have to decide the best one using some strategies. The NP-Hard problem is something where time is unknown, and it is not a decision problem like NP-Complete. However, in quantum computing the NP-Complete problem can be solved using Grover's algorithm. A normal algorithm to solve an NP-Complete problem will take $O(n^2)$ to draw a decision; however, Grover's algorithm will take 0 (SquareRoot(n)).

The NP-Hard problems are not decision problems, but they are optimization problems. An NP-Hard can be solved if, and only if, the NP-Hard problem can be reduced into an NP-Complete problem. Since an NP-Hard problem does not yield decisions, the time complexity is unknown. The best examples are the halting problem, the traveling salesman problem and so on. Quantum computers work well in large search spaces and optimize objective functions by placing the input in various states in Hilbert spaces. Optimization problems should follow the QUBO schema to be solved efficiently. This involves obtaining multiple solutions and introducing slack variables to penalize a solution when the constraints are violated, and trying to reduce the penalty to obtain the best optimal solution.

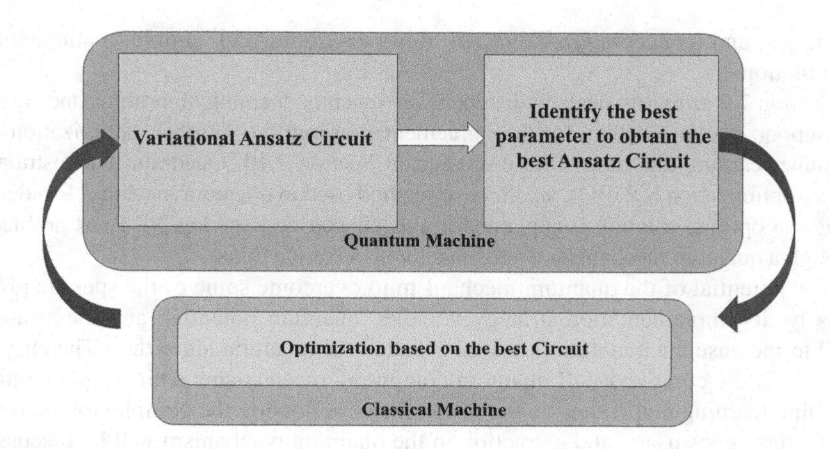

FIGURE 7.9 The variational quantum algorithm framework.

To solve an NP-Hard problem in quantum machines, variational quantum algorithms have been widely used. The variational quantum algorithm is shown in Figure 7.9.

Variational quantum algorithms are adaptive in nature and optimization will be carried out on the fly. Variational circuits are termed as ansatz circuits, where multiple gates have been attached to the input transmissions, and parameters associated with the functions are variational. The problem is defined as a parametrized objective function, which defines the feature space. The quantum circuit identifies the parameters for optimizing the objective function, which can be further used to fine-tune the solutions present in classical machines. Variational quantum algorithms have been broadly classified as two categories of algorithms namely: quantum approximate optimization algorithms and variational quantum eigen solvers. Although NP-Hard problems have been solved using quantum computers, the complexity has little variation when compared with computation with classical computers, since it is still at the initial level.

7.11 SUMMARY

This chapter has provided an overview of the utilization of quantum computing in the machine learning domain. It began by discussing the complexity of quantum computing and detailing quantum computing with figures. The representation of quantum states as qubits and the superpositioning of qubits in the bloch sphere are detailed with appropriate representations. Entanglement is the main critical part of quantum computing, detailed in the feature map section for better understanding.

Sections 7.1 and 7.2 cover complexity, superposition, and entanglement, then Section 7.3 details machine learning algorithm implementation in quantum computing. Section 7.4 deals with quantum principal component analysis, an effective dimensionality reduction, and a potential feature extraction in machine learning. In Section 7.5, linear systems implementation on quantum machines is depicted through mathematical background and is conceptualized using figures. Similarly, Sections 7.6, 7.7, and 7.8 detail unsupervised learning techniques implementation on quantum machines and conceptualize through flow diagrams and simple mathematical formulations.

Section 7.9 primarily deals with supervised machine learning algorithms and selects the support vector machine for the implementation purpose. Further, optimization for machine learning algorithms are discussed in Section 7.10. Quadratic unconstrained binary optimization (QUBO), an effective method used in quantum machines for identifying the optimal solution is explained in this chapter and solving NP-Hard problems through a quantum mechanism is illustrated with a simple figure.

The potential of the quantum mechanism to overcome some of the specific problems by its implementation strategy is called quantum potential, and the strategy used to increase the quantum potential is called the quantum algorithm. The chapter deals with the complexity of quantum computing mechanisms while implementing machine learning algorithms as the prima facie. Following the complexity analysis, the feature maps usage and extraction in the quantum mechanism will be discussed for a clear understanding.

This chapter relates in brief the concepts of quantum mechanisms in machine learning through simple language for the better understanding of learners. The chapter aims to conceptualize the implementations and mechanisms through simple diagrams, flow diagrams, and using straightforward mathematics for clarity. Furthermore, explaining the subject from conceptualizing to depth through simple mathematics makes it accessible for beginners.

For practice and simulating the quantum mechanisms, practitioners are recommended to access https://qiskit.org/

REFERENCES

1 Hirvensalo, M. (2003). *Quantum computing*. Springer Science & Business Media.

2 National Academies of Sciences, Engineering, and Medicine. (2019). *Quantum computing: Progress and prospects*. The National Academies Press. https://doi.org/10.17226/25196

3 Williams, C. P., Clearwater, S. H., & others. (1998). *Explorations in quantum computing*. Springer.

4 Hey, T. (1999). Quantum computing: An introduction. *Computing & Control Engineering Journal*, *10*(3), 105–112.

5 Bennett, C. H., Bernstein, E., Brassard, G., & Vazirani, U. (1997). Strengths and weaknesses of quantum computing. *SIAM Journal on Computing*, *26*(5), 1510–1523.

6 Schuld, M., & Killoran, N. (2019). Quantum machine learning in feature Hilbert spaces. *Physical Review Letters*, *122*(4), 40504.

7 Córcoles, A. D., Kandala, A., Javadi-Abhari, A., McClure, D. T., Cross, A. W., Temme, K., Nation, P. D., Steffen, M., & Gambetta, J. M. (2019). Challenges and opportunities of near-term quantum computing systems. *Proceedings of the IEEE*, *108*(8), 1338–1352.

8 Schuld, M., & Petruccione, F. (2021). Representing data on a quantum computer. *Machine Learning with Quantum Computers*, *1*, 147–176.

9 Goto, T., Tran, Q. H., & Nakajima, K. (2021). Universal approximation property of quantum machine learning models in quantum-enhanced feature spaces. *Physical Review Letters*, *127*(9), 90506.

10 Goto, T., Tran, Q. H., & Nakajima, K. (2020). Universal approximation property of quantum feature map. *ArXiv Preprint ArXiv:2009.00298*.

11 Giannakis, D., Ourmazd, A., Pfeffer, P., Schumacher, J., & Slawinska, J. (2022). Embedding classical dynamics in a quantum computer. *Physical Review A*, *105*(5), 52404.
12 Thumwanit, N., Lortaraprasert, C., & Raymond, R. (2021). Trainable discrete feature embeddings for quantum machine learning. *2021 58th ACM/IEEE Design Automation Conference (DAC)*, 1352–1355.
13 Yang, R., Bosch, S., Kiani, B., Lloyd, S., & Lupascu, A. (2023). Analog quantum variational embedding classifier. *Physical Review Applied*, *19*(5), 54023.
14 Lloyd, S., Mohseni, M., & Rebentrost, P. (2014). Quantum principal component analysis. *Nature Physics*, *10*(9), 631–633.
15 Parmar, J. B. (2022). *Quantum principal component analysis*. The National Academies Press. https://doi.org/10.13140/RG.2.2.14437.06885
16 He, C., Li, J., Liu, W., Peng, J., & Wang, Z. J. (2022). A low-complexity quantum principal component analysis algorithm. *IEEE Transactions on Quantum Engineering*, *3*, 1–13.
17 Al-Alimi, D., Al-Qaness, M. A. A., Cai, Z., Dahou, A., Shao, Y., & Issaka, S. (2022). Meta-learner hybrid models to classify hyperspectral images. *Remote Sensing*, *14*(4), 1038.
18 Petersen, I. R. (2016). Quantum linear systems theory. *ArXiv Preprint ArXiv:1603.04950*.
19 Harrow, A. W., Hassidim, A., & Lloyd, S. (2009). Quantum algorithm for linear systems of equations. *Physical Review Letters*, *103*(15), 150502.
20 Kerenidis, I., & Prakash, A. (2020). Quantum gradient descent for linear systems and least squares. *Physical Review A*, *101*(2), 22316.
21 Maier, T., Jarrell, M., Pruschke, T., & Hettler, M. H. (2005). Quantum cluster theories. *Reviews of Modern Physics*, *77*(3), 1027.
22 Horn, D., & Gottlieb, A. (2001). The method of quantum clustering. *Advances in Neural Information Processing Systems*, *14*, 1–7
23 Poggiali, A., Berti, A., Bernasconi, A., Del Corso, G. M., & Giudotti, R. (2022). Clustering classical data with quantum K-means. *Proceedings of the 23rd Italian Conference on Theoretical Computer Science*, Roma, Italy, 7–9.
24 Wu, Z., Song, T., & Zhang, Y. (2022). Quantum k-means algorithm based on Manhattan distance. *Quantum Information Processing*, *21*(1), 19.
25 Benlamine, K., Bennani, Y., Grozavu, N., & Matei, B. (2020). Quantum collaborative k-means. *2020 International Joint Conference on Neural Networks (IJCNN)*, 1–7.
26 Voevodski, K. (2021). Large scale k-median clustering for stable clustering instances. *International Conference on Artificial Intelligence and Statistics*, *1*, 2890–2898.
27 Aïmeur, E., Brassard, G., & Gambs, S. (2007). Quantum clustering algorithms. *Proceedings of the 24th International Conference on Machine Learning*, 1–8.
28 Kumar, V., Bass, G., Tomlin, C., & Dulny, J. (2018). Quantum annealing for combinatorial clustering. *Quantum Information Processing*, *17*, 1–14.
29 Wei, A. Y., Naik, P., Harrow, A. W., & Thaler, J. (2020). Quantum algorithms for jet clustering. *Physical Review D*, *101*(9), 94015.
30 Duan, B., Yuan, J., Yu, C.-H., Huang, J., & Hsieh, C.-Y. (2020). A survey on HHL algorithm: From theory to application in quantum machine learning. *Physics Letters A*, *384*(24), 126595.
31 Liu, X., Xie, H., Liu, Z., & Zhao, C. (2022). Survey on the improvement and application of HHL algorithm. *Journal of Physics: Conference Series*, *2333*(1), 12023.
32 Glover, F., Kochenberger, G., & Du, Y. (2018). A tutorial on formulating and using QUBO models. *ArXiv Preprint ArXiv:1811.11538*.
33 Papalitsas, C., Andronikos, T., Giannakis, K., Theocharopoulou, G., & Fanarioti, S. (2019). A QUBO model for the traveling salesman problem with time windows. *Algorithms*, *12*(11), 224.

34 Date, P., Arthur, D., & Pusey-Nazzaro, L. (2021). QUBO formulations for training machine learning models. *Scientific Reports*, *11*(1), 10029.
35 Bologna, G., & Hayashi, Y. (2015). Qsvm: A support vector machine for rule extraction. *Advances in Computational Intelligence: 13th International Work-Conference on Artificial Neural Networks, IWANN 2015*, Palma de Mallorca, Spain, June 1012, 2015. Proceedings, Part II 13, 276–289.
36 Ahmed, S. (2019). *Pattern recognition with Quantum Support Vector Machine (QSVM) on near term quantum processors*. Brac University.
37 Shan, Z., Guo, J., Ding, X., Zhou, X., Wang, J., Lian, H., Gao, Y., Zhao, B., & Xu, J. (2022). *Demonstration of breast cancer detection using QSVM on IBM quantum processors*. Research Square.

8 Quantum Classification

Durgadevi Palani and Akila Krishnamoorthy
SRM Institute of Science and Technology, India

Quantum classification is a subfield of quantum machine learning that focuses on developing algorithms and techniques to classify data using quantum computing principles. It aims to leverage the power of quantum computing to process and analyze data in ways that could potentially outperform classical machine learning algorithms.

In quantum classification, quantum states are used to represent and process data, and quantum algorithms are designed to perform classification tasks, as shown in Figure 8.1. Quantum classifiers can be trained on classical data or quantum data, and they exploit quantum properties like superposition and entanglement to enhance the learning and classification process.

There are several approaches to quantum classification, including:

1. **Quantum Variation Algorithms:** These algorithms utilize quantum circuits that can be trained to optimize a cost function, similar to classical neural networks. Examples include the quantum variation classifier (QVC) and the quantum neural network (QNN).
2. **Quantum Support Vector Machines (QSVM):** QSVM is a quantum version of the classical support vector machine (SVM) algorithm. It employs a quantum kernel to map data into a high-dimensional quantum feature space, enabling classification based on quantum measurements.
3. **Quantum k-nearest neighbors (QK-NN):** This is a quantum extension of the classical k-nearest neighbor's algorithm. It involves encoding classical data into quantum states and using quantum distance measures to determine the nearest neighbors.
4. **Quantum Decision Trees:** Quantum decision trees are quantum versions of classical decision trees. They utilize quantum circuits to perform splitting and decision-making processes.

It is important to note that quantum classification is an active area of research, and the field is still in its early stages. Many of the proposed algorithms are still being developed and optimized, and their performance is being evaluated on various quantum hardware platforms. Additionally, quantum machine learning algorithms often rely on classical data pre-processing steps and classical machine learning techniques in combination with quantum components. Hybrid quantum-classical approaches are commonly used to bridge the gap between classical and quantum computing and take advantage of the strengths of both paradigms.

DOI: 10.1201/9781003429654-11

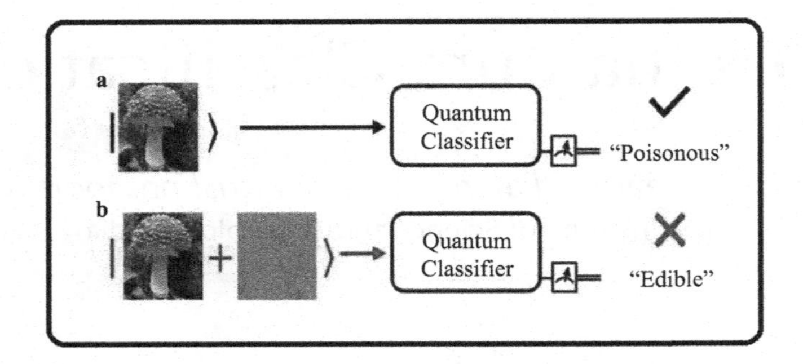

FIGURE 8.1 Quantum classification.

8.1 NEAREST NEIGHBORS

Nearest neighbors algorithms are widely used in classical machine learning for classification tasks based on features, as depicted in Figure 8.2. These algorithms make predictions by finding the closest training examples (neighbors) to a given input data point and using their labels or values to make predictions. In the context of quantum classification, the application of nearest neighbor's techniques shows promise in leveraging the power of quantum computing.

This section explores the adaptation of nearest neighbor's algorithms to quantum data, covering data encoding, distance measurement, nearest neighbor selection, and classification using quantum nearest neighbors.

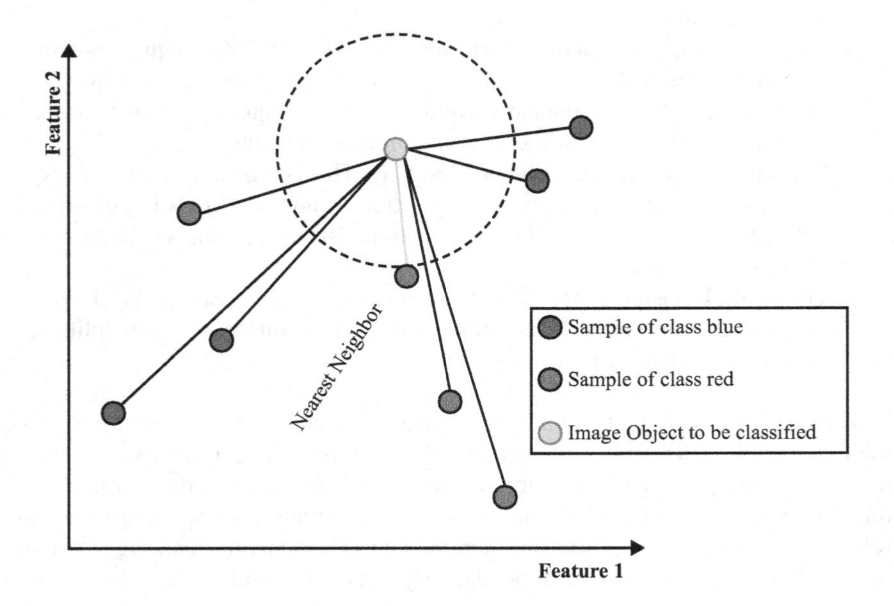

FIGURE 8.2 Nearest neighbors algorithms.

8.1.1 Data Encoding

In quantum classification, classical data needs to be encoded into quantum states. One common approach is amplitude encoding, where the amplitudes of a quantum state represent the classical features.

For example, given a classical feature vector $v = (v\_1, v\_2, \ldots, v\_n)$, the quantum state can be represented as

$$|v\rangle = \Sigma\_iv\_i|i\rangle \tag{8.1}$$

8.1.2 Distance Measurement

The quantification of similarity or dissimilarity between quantum states is achieved through the utilization of distance metrics. Among these metrics, the fidelity is a widely employed measure, which is defined as

$$\mathrm{F}(\rho 1, \sigma 1) = \mathrm{Tr}\left(\sqrt{}\left(\sqrt{}\,\rho 1 \sigma 1 \sqrt{}\,\rho 1\right)\right) \tag{8.2}$$

where Tr denotes the trace operation.

Another distance measure is the quantum state overlap, given by the inner product of the two quantum states,

$$\langle \rho | \sigma \rangle = \mathrm{Tr}(\rho \dagger \sigma) \tag{8.3}$$

8.1.3 Nearest Neighbor Selection

To determine the nearest neighbors of a test quantum state, the distances between the test state and each training state are calculated using the chosen distance metric. Let $|\psi\_\text{test}\rangle$ and $|\psi\_i\rangle$ represent the test quantum state and the i-th training quantum state, respectively. The distance between them can be computed using the fidelity or quantum state overlap measures discussed earlier.

8.1.4 Classification

Once the distances are calculated, the k-nearest neighbors with the smallest distances are selected. To determine the class label of a test quantum state, the class labels of its nearest neighbors are utilized. This can be achieved through various decision rules, such as majority voting or weighted voting based on the distances. The final classification decision is based on the class labels of the nearest neighbors.

Equations:

Amplitude Encoding: Given a classical feature vector $v = (v\_1, v\_2\ldots v\_n)$, the quantum state can be represented as $|v\rangle = \Sigma\_iv\_i|i\rangle$.

Fidelity: Fidelity among two quantum states $\rho 1$ and $\sigma 1$ which is defined as $F(\rho 1, \sigma 1) = \mathrm{Tr}(\sqrt{}(\sqrt{}\rho 1 \sigma 1 \sqrt{}\rho 1))$.

Quantum State Overlap: The quantum state overlap among two quantum states $\rho 1$ and $\sigma 1$which is represented by the inner product $\langle \rho 1 | \sigma 1 \rangle = \mathrm{Tr}(\rho 1 \dagger \sigma 1)$.

8.2 SUPPORT VECTOR MACHINES THROUGH GROVER'S SEARCH ALGORITHM

Support vector machines (SVMs) are powerful classifiers widely used in classical machine learning. In the realm of quantum machine learning, researchers have explored the integration of Grover's search algorithm with SVMs to enhance their performance.

8.2.1 Support Vector Machines in Classical Machine Learning

In classical SVMs, a hyperplane is created in a high-dimensional feature space to separate data points belonging to different classes. Support vectors, refers to the data points which are closest to the hyperplane, and have a substantial role in defining the decision boundary. The training process of an SVM involves solving a quadratic programming problem to identify the optimal hyperplane.

8.2.2 Unlocking Quantum Speedup in Search Problems: Grover's Search Algorithm

Grover's search is one of the quantum algorithms known for its ability to perform unstructured search tasks with a quadratic speedup over classical algorithms.

Grover's search algorithm is a quantum algorithm that offers a quadratic speedup which is compared to classical algorithms for unstructured examine problems. Grover's search algorithm, introduced by Lov Grover in 1996, has found applications in diverse fields such as database search and optimization problems. Figure 8.3 illustrates the high-level steps of Grover's search algorithm.

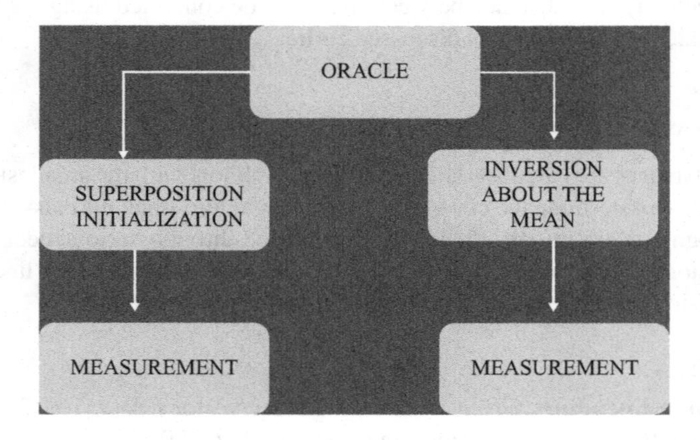

FIGURE 8.3 Steps involved in Grover's algorithm.

The algorithm consists of four main steps: superposition initialization, oracle application, inversion about the mean, and measurement.

1. **Superposition initialization:** The algorithm starts by preparing the quantum state in a superposition of all possible inputs. This is achieved by applying a Hadamard gate to each qubit in the quantum register, an equal superposition, which encompasses all possible states simultaneously.
2. **Oracle application:** The oracle is a quantum circuit that marks the desired solution(s) in the search space. It provides a phase inversion for the target state(s) while leaving the other states unchanged. The oracle acts as a black box, implementing a specific function related to the search problem.
3. **Inversion about the mean:** After applying the oracle, the algorithm performs an inversion about the mean operation. This step involves reflecting the amplitudes of the quantum state about their mean, effectively amplifying the amplitude of the marked solution(s) and suppressing the others. It is achieved by applying a sequence of quantum gates, including the Hadamard gate and the phase gate.
4. **Measurement:** Finally, the algorithm performs a measurement of the quantum state. The measurement collapses the superposition into a single classical state, providing the desired solution(s) with high probability.

Grover's search algorithm iteratively repeats steps 2 and 3 a certain number of times to increase the likelihood of measuring the desired solution(s) in the final measurement step. The number of the iterations required is reliant on size of the search space.

It can be used to enhance SVMs by accelerating the computation of the kernel matrix, a crucial step in training SVMs. The integration of Grover's search with SVMs involves utilizing the quantum algorithm to search for the support vectors more efficiently. This integration can reduce the time complexity of training SVMs, making them more efficient in large-scale classification problems. The hyperplane equation in classical SVMs is specified by:

$$F(x1) = \text{sign}\left(\Sigma\_i\alpha 1\_iy1\_iK(x1, x1\_i) + b\right) \tag{8.4}$$

where $\alpha 1\_i$ is the Lagrange multiplier, $y1\_i$ is the class label, $K(x1, x1\_i)$ is the kernel function, and b denotes bias term.

Quantum Kernel Matrix Computation:

The quantum kernel matrix can be computed using Grover's search algorithm as follows:

$$|K\rangle = \Sigma\_i\sqrt{(\alpha\_i)}|x\_i\rangle|y\_i\rangle \tag{8.5}$$

where $|x\_i\rangle$ represents the quantum state encoding the feature vector x_i and $|y\_i\rangle$ is the quantum state encoding the class label y_i.

8.2.3 Advantages of Grover's Search with SVMs

- **Speedup in training time:** Grover's search can accelerate the computation of the kernel matrix, reducing the time complexity of SVM training.
- **Improved scalability:** The integration of Grover's search can enable SVMs to handle larger datasets more efficiently.
- **Potential for enhanced performance:** The efficient computation of the kernel matrix can lead to improved generalization and classification accuracy.

8.3 SUPPORT VECTOR MACHINES WITH EXPONENTIAL SPEEDUP

Quantum computing offers the potential for exponential speedup in solving certain computational problems. In the context of SVMs, quantum algorithms can significantly reduce the computational complexity of training and prediction tasks, leading to exponential speedup.

Two notable approaches for achieving exponential speedup in SVMs are quantum matrix inversion and quantum feature maps.

Quantum matrix inversion leverages quantum algorithms, such as the HHL (Harrow-Hassidim-Lloyd) algorithm, to efficiently invert the kernel matrix. This exponential speedup reduces the computational complexity of SVM training. The equation for quantum matrix inversion using the HHL algorithm is:

$$\mathrm{A}|x\rangle = |b\rangle \tag{8.6}$$

where A is the matrix to be inverted, $|x\rangle$ is the solution vector, and $|b\rangle$ is the input vector.

Quantum feature maps are instrumental in support vector machines (SVMs) as they facilitate the transformation of input data into a higher-dimensional feature space, enabling the separation of classes using a linear hyperplane. Quantum feature maps leverage quantum algorithms, such as the quantum kernel trick, to efficiently perform feature mapping. This results in a significant reduction in the computational resources required for SVM training, leading to exponential speedup. The equation for quantum feature mapping is given by:

$$K(x1, y1) = \langle \phi(x1)/\phi(y1) \rangle \tag{8.7}$$

where $\phi(x1)$ and $\phi(y1)$ are the quantum feature representations of input vectors $x1$ and $y1$, respectively, and $K(x1, y1)$ is the kernel function.

The integration of quantum algorithms with exponential speedup into SVMs offers several implications for quantum classification. First, it enables enhanced scalability, allowing SVMs to handle larger datasets more efficiently. The reduction in computational complexity enables SVMs to tackle problems that were previously computationally intractable using classical approaches. Second, exponential speedup leads to faster training and prediction times, which is particularly advantageous for real-time and time-sensitive applications. Finally, the computational advantages of quantum SVMs can improve classification accuracy by allowing SVMs to learn more complex decision boundaries and capture intricate patterns in the data.

However, there are challenges to address in realizing the full potential of SVMs with exponential speedup. Quantum hardware constraints, such as the need for a sufficient number of qubits and reliable quantum gates, pose implementation challenges. Further research is needed to develop efficient quantum algorithms specifically designed for SVMs, optimizing them for various problem types and improving their performance on quantum hardware. Additionally, translating theoretical quantum algorithms into practical quantum circuits requires addressing considerations such as circuit depth, connectivity constraints, and the integration of error correction techniques.

8.4 COMPUTATIONAL COMPLEXITY

The computational complexity of quantum classification algorithms can be analyzed in terms of the number of quantum operations or qubits required for the algorithm. In general, quantum algorithms can offer exponential speedup by solving certain problems in a time that scales exponentially better than classical algorithms. To illustrate, the comparison between classical and quantum algorithms in terms of computational complexity is represented in Table 8.1.

In Table 8.1, "N" represents total of data points, "d" represents the dimensionality of the feature space, and "M" represents total of trees in random forests. The complexity of classical classification algorithms, such as SVMs, decision trees, and random forests, scales polynomially with the dataset size and feature dimensionality.

In Table 8.2, the complexity of quantum classification algorithms, such as quantum SVMs, quantum decision trees, and quantum random forests, is presented. The quantum complexity for these algorithms can be affected by the number of data points, the feature dimensionality, and the number of trees in the case of random forests. These quantum algorithms offer potential exponential speedup over classical approaches.

TABLE 8.1
Computational Complexity of Conventional Algorithms

Algorithm	Complexity
SVM	$\exp(O(\log(N) + \log(d)))$
Decision trees	$\exp(O(\log(N) + \log(d) + \log(\log(N))))$
Random forests	$\exp(O(\log(M) + \log(N) + \log(d) + \log(\log(N))))$

TABLE 8.2
Computational Complexity of Quantum Classification Algorithms

Algorithm	Complexity
Quantum SVM	$\exp(O(\log(N\char`^3)))$
Quantum Decision Trees	$\exp(O(\log(N) + 2\log(N) + \log(d) + \log(\log(N))))$
Quantum Random Forests	$\exp(O(\log(M) + 2\log(N) + \log(d) + \log(\log(N))))$

TABLE 8.3
Comparison of Computational Complexity

Algorithm	Classical Complexity	Quantum Complexity
SVM	$\exp(O(\log(N) + \log(d)))$	$\exp(O(\log(N^3)))$
Decision trees	$\exp(O(\log(N) + \log(d) + \log(\log(N))))$	$\exp(O(\log(N) + 2\log(N) + \log(d) + \log(\log(N))))$
Random forests	$\exp(O(\log(M) + \log(N) + \log(d) + \log(\log(N))))$	$\exp(O(\log(M) + 2\log(N) + \log(d) + \log(\log(N))))$

Table 8.3 provides a comparison of the computational complexity between classical and quantum algorithms for SVMs, decision trees, and random forests. It highlights the significant reduction in complexity achieved by quantum algorithms, which can potentially lead to exponential speedup.

However, it is important to note that achieving this exponential speedup in practice requires overcoming several challenges. One of the primary challenges is the implementation of quantum algorithms on physical quantum hardware. Quantum computers are still in their early stages of development, and current quantum systems have limited qubit coherence times and high error rates. As a result, implementing quantum classification algorithms on existing quantum hardware can be challenging.

Moreover, the complexity of quantum classification algorithms depends on the specific problem and the size of the dataset. While quantum algorithms can offer exponential speedup for certain tasks, they may still require a significant number of qubits or quantum gates to achieve this advantage. As the size of the dataset increases, the number of required qubits and quantum operations also grows, which can pose practical limitations on the scalability of quantum classification algorithms.

8.5 SUMMARY

In summary, quantum classification algorithms have the potential to provide exponential speedup over classical approaches, enabling faster and more accurate classification. The comparison tables demonstrate the significant reduction in computational complexity achieved by quantum algorithms. However, realizing the full potential of quantum classification in practice requires addressing challenges related to quantum hardware limitations and optimizing the scalability and efficiency of quantum algorithms.

FURTHER READING

1. Schuld, M., Sinayskiy, I., & Petruccione, F. (2015). An introduction to quantum machine learning. *Contemporary Physics*, 56(2), 172–185.
2. Biamonte, J., & Wittek, P. (2017). Quantum machine learning. *Nature*, 549(7671), 195–202.
3. Cong, I., Duan, R., & Fruchtman, A. (2019). Quantum-inspired machine learning: Concepts and recent advances. *Frontiers of Physics*, 14(5), 53602.

4. Schuld, M., Fingerhuth, M., & Petruccione, F. (2018). Implementing a distance-based classifier with a quantum interference circuit. *EPL (Europhysics Letters)*, 119(6), 60002.
5. Havlíček, V., Córcoles, A. D., Temme, K., Kandala, A., Chow, J. M., Gambetta, J. M., … Gambetta, J. M. (2019). Supervised learning with quantum-enhanced feature spaces. *Nature*, 567(7747), 209–212.
6. Cai, X. D., Wu, D., Su, Z. E., Chen, M. C., Wang, X. L., Li, L., … Pan, J. W. (2015). Experimental quantum computing to solve systems of linear equations. *Physical Review Letters*, 114(11), 110504.
7. Rebentrost, P., Mohseni, M., & Lloyd, S. (2014). Quantum support vector machine for big data classification. *Physical Review Letters*, 113(13), 130503.
8. Liu, L., Wang, J., Chen, L., & Zhu, S. (2020). Quantum k-nearest neighbor algorithm for classical data. *Quantum Information Processing*, 19(9), 1–15.
9. Benedetti, M., Realpe-Gómez, J., Biswas, R., & Perdomo-Ortiz, A. (2019). Quantum-assisted learning of graphical models with arbitrary pairwise connectivity. *Quantum Science and Technology*, 4(2), 024010.
10. Schuld, M., Bergholm, V., Gogolin, C., & Izaac, J. (2019). Evaluating analytic gradients on quantum hardware. *Physical Review A*, 99(3), 032331.

9 Boosting in QML

Ponnuviji Namakkal Ponnusamy
RMK College of Engineering and Technology, India

Indra Priyadharshini Sundar
VIT-Chennai, India

Nirmala Ganapath
RMD Engineering College, India

9.1 INTRODUCTION TO BOOSTING

With the ability to solve computational problems that are impossible for traditional machines, quantum computing has the capacity to revolutionize computing and make a significant impact in key industries such as agriculture, finance, energy, and design of materials science. Design choices are being made to try to squeeze the most processing possible out of a machine when quantum computing systems with between 50 and 100 qubits and larger are built. This has led to research into simple quantum volume (SQV) [1], which can be modeled by multiplying a machine's computational qubit count by the number of gates anticipated as being able to execute error-free, as shown in Figure 9.1. The smallest unit of storage in quantum computing is qubits, that is, physical quantum bits which are exceedingly error-prone. This implies that computing on these machines is restricted by the quick expiring lifespan of qubits. This is one constraint on SQV at the moment. To amend this, system designers are working to create better physical qubits, however, this is very challenging, so classical systems may be recommended for use to lighten the load.

In particular, quantum error correction is a traditional control method to reduce the rate of mistakes in qubits and improves the SQV. The next step in error correction is to encode a group of logical qubits that will be fed through algorithms into a group of physically faulty qubits. A particular quantum circuit, known as a syndrome, extracts data about the device's current state without interfering with the underlying process. The error-correcting protocol can be used to convert the data into a set of corrections through a process known as decoding. If the corrections are chosen correctly, the system should be put back into the proper logical state. The SQV can be rapidly expanded by fully fault-tolerant machines by reducing qubit errors exponentially with the code distance.

DOI: 10.1201/9781003429654-12

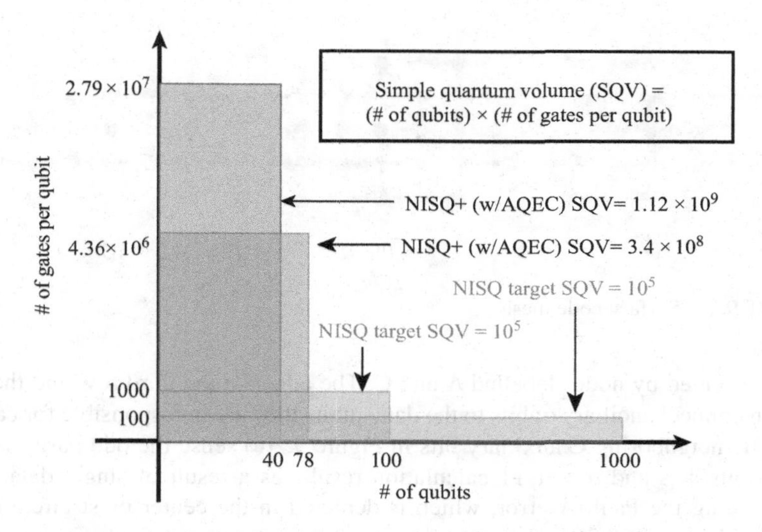

FIGURE 9.1 Quantum computer boosting with error correction codes.

9.1.1 Quantum Error Correction

Due to their intrinsic fragility, qubits must be kept isolated from outside influences in order to retain their value. Decoherence, such as the quantum state's transition from the general state $|S = \beta|0 + \gamma|1$ to the ground state $|S = |0$ occurs quickly in many different types of actual physical qubits, frequently on the magnitude order of tens of nanoseconds. As a result, algorithms are severely limited in their ability to run successfully for long periods of time without any sort of system modifications.

Quantum error correction procedures have been created to tackle this. These involve encoding a smaller number of logical qubits, which are used for computation algorithms, into a larger number of physical qubits, increasing reliability. Generally speaking, creating quantum error correction methods is challenging since doing so will cause the data in the system's qubits to be destroyed. Protocols rely on adding additional qubits that work together with the main set of qubits and are measured in order to acquire information about errors indirectly. The locations of data qubits which include errors are subsequently inferred using the measurement data.

Physical qubits are susceptible to Surface Code Errors because they are mathematically represented by two complex coefficients that have a continuous range of possible values. The surface code's use of quantum mechanics has the advantage that these constant faults can be eventually broken into a small number of distinct errors. More specifically, the surface code's operation converts these continuous constant faults occurring on the data into Pauli error operators, which take the form {I, A, B, C}. One of the key components of the code that makes it possible for error detection and correction to take place is this. A two-dimensional error-correcting code is used in the surface code technique to provide error discretization, detection, and correction. This is shown in Figure 9.2.

Data qubits are represented by second and fourth row circles, leaving the alternate circles for ancillary qubits calculating the A and C stabilizers correspondingly, which

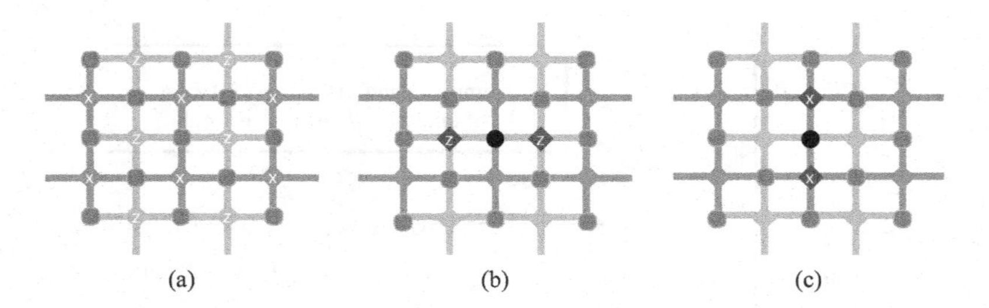

FIGURE 9.2 Surface code mesh.

are represented by nodes labelled A and C. The edges in the third row and the third column connect ancillary qubits to the data qubits they are in responsible for calculation. The neighboring C auxiliary bits in Figure 9.2(c) sense the odd parity in their data qubits sets and report +1 calculation results as a result of single data qubits experiencing the Pauli-A error, which is depicted in the center most circle on the diagram. In Figure 9.2(c), the center most circle encounters a Pauli-C error, which cause the A auxiliary qubits that are vertically adjacent to it to report measurement values of 1.

An error syndrome is created as an outcome of ancillary qubits' interactions with all data qubits in their immediate vicinity and subsequent measurements. This sequence of operations, among which every supplementary qubit involves a four-qubit operator known as a stabilizer, makes up the stabilizer circuit.

9.1.2 Machine Learning for Boosting

Due to its popularity in real-world applications, machine learning (ML) has drawn a lot of interest in the past ten years. Given the wide range of uses for ML, there has been a lot of interest in figuring out specific learning tasks that quantum computers will be able to accelerate. For accurate practical machine learning applications, there has been a rush of quantum algorithms that potentially offer exponential or polynomial quantum speedups over classical computers. Theoretical research on quantum machine learning (QML) formerly focused on creating effective quantum algorithms with desirable quantum complexity to address engaging learning challenges. More recently, efforts have been made to comprehend how noisy tiny quantum devices interact with quantum machine learning algorithms. QML provides us with various algorithms for implementing quantum learning as well as classical learning tasks such as: (i) transforming classical machine learning algorithms like SVM, linear algebra, classification algorithms based on kernel, gradient computing algorithms, perceptron learning algorithms, and clustering algorithms into quantum efficient algorithms; (ii) handling the arbitrary quantum states in the PAC setting, shadow tomography of quantum states, and learning of quantum objects like the classical class of stabilizer states, low-entanglement states; (iii) learning Boolean-valued concept classes through quantum framework; (iv) optimization quantum algorithms; (v) quantum machine learning algorithms for generative models, and so forth.

Assume A is a QML algorithm with a theoretically highly efficient performance. However, the performance of A is poor when used with a noisy quantum computer, meaning that only slightly more than half of the inputs result in the right output. Can we make A perform better so that its output is accurate on two-thirds of the inputs? A traditional adaptive boosting method, generally known as AdaBoost, can be employed right away to transform a weak quantum learning algorithm into a strong one. There are quantum boosting techniques available that quadratically outperform the traditional AdaBoost algorithm. We can even convert a poor and imprecise QML into a faster algorithm that can run quadratically faster using the quantum boosting technique. If we have a Boolean concept class C and use a weak learner with accuracy γ, which requires time R(C) for training, then the time complexity of the AdaBoost algorithm scales as VC(C)·poly(R(C), $1/\gamma$). Here, VC(C) represents the VC-dimension of concept class C, which measures its capacity to fit data, and poly() denotes a polynomial function that accounts for the combined influence of the VC-dimension and the weak learner's training time.

9.2 CLASSICAL BOOSTING

Many modern statistical and machine learning methods heavily rely on optimizing an objective or loss function. For example, in simple linear regression, the goal is to find a weight vector w that provides a good approximation of the target value *yi* for each input *xi*, given a set of examples (*x*1, *y*1), …, (*xm*, *ym*), where $xi \in Rn$ and $yi \in R$. The objective is to minimize the average (or sum) of the squared errors, which involves finding the optimal values for the elements of $w \in Rn$.

$$L(w) = \frac{1}{m} \sum\nolimits_{i=1}^{m} (w.xi - yi)^n$$

In various machine learning techniques, such as linear regression, the loss function is crucial. This function, often a squared or quadratic loss in this context, quantifies the squared error for each component $(w\ xi - yi)^2$. The primary objective is to minimize the average of these losses across all *m* instances. This optimization process can also be applied to other methods like neural networks, support vector machines, logistic regression, maximum likelihood, and more. Defining a clear loss function and minimizing it provides a well-defined learning goal. Additionally, the flexibility to modify objective functions makes it adaptable for tackling new and diverse learning challenges.

9.2.1 AdaBoost

Originally AdaBoost was not designed as a procedure for optimizing objective function. Research has shown that AdaBoost turns out to greedily minimize the loss function, which helps us in understanding and extending the algorithm. Therefore, AdaBoost can be generalized for working with loss functions and boost performance.

Freund and Schapire introduced the remarkable boosting algorithm called AdaBoost, which addresses the question of whether a weak learner, treated as a black box, can be utilized to create a strong learner. Their AdaBoost method, known for its simplicity and practicality, has been a significant breakthrough with numerous applications in game theory, statistics, optimization, vision, speech recognition, and information geometry. In recognition of the algorithm's effectiveness in both theory and practice, Freund and Schapire were honored with the Godel Prize in 2003 [2].

The loss of interest for a classifier H on a labeled example (x, y) is given by $1\{H(x) \neq y\_i\}$, where:

- $H(x)$ represents the prediction of classifier H on input x.
- y_i is the true label (ground truth) of the example x.
- $1\{\}$ is an indicator function that takes the value 1 if the condition inside the curly braces is true, and 0 otherwise.

In simple terms, the loss of interest is equal to 1 if the classifier H misclassifies the example (x, y), indicating that its prediction does not match the true label y_i. If the classifier makes the correct prediction, the loss of interest is 0, as there is no error in this case. This loss function is commonly used in classification tasks to evaluate the performance of a classifier and guide the learning process to minimize misclassifications.

The AdaBoost Training error can be defined as

$$\frac{1}{m}\sum_{i=1}^{m} 1\{H\left(xi\right) \neq yi$$

where $(x_1, y_1), \ldots, (x_m, y_m)$ is the training set.

H is the combined classifier, $H(x) = \text{sign}(F(x))$, and $F(x) = \sum_{t=1}^{T} atht\left(x\right)$.

9.2.1.1 How Does AdaBoost Work?

Assume A is a weak PAC (probably approximately correct) learning algorithm for C that runs in time R(C) with bias $\gamma > 0$. That is, the algorithm performs similarly or slightly better than the random guessing technique. The objective of boosting is as follows.

For every unknown distribution N: $\{0, 1\}\ n \rightarrow [0, 1]$ and unknown concept $c \in$ C, construct a hypothesis H that satisfies Prx $\sim N[H(x) = c(x)] \geq 2/3$. The AdaBoost algorithm by Freund and Schapire generates such a classifier H by invoking the algorithm A polynomially many times.

The AdaBoost algorithm begins by obtaining M different labeled examples in the form of a set $S = \{(x_i, c(x_i)): i \in [M]\}$, where x_i belongs to N.

Next, the algorithm iterates T steps (for some M). At each step, it defines a distribution N_t based on the previous distribution N_{t-1} and invokes a weak learner A on the sample S and the current distribution N_t.

The output hypothesis h_t of the weak learner A is then used to calculate the weighted error $\varepsilon_t = \Pr(x \sim N_t)[h_t(x) \neq c(x)]$, which represents the probability of h_t

1. Input: M different labelled examples $S = \{(X_i, c(X_i)): i \in [M]\}$ where x_i belongs to N
2. Initialize distribution N_1 as the uniform distribution on S
3. For t = 1 to T:
   a. Invoke weak learner A on sample S and distribution N_t to obtain hypothesis h_{tv}
   b. Calculate weighted error ε_t $Pr(x \sim N_t)[h_t(x) \neq c(x)]$
   c. Compute weight $\alpha_t = 1/2 \text{ In } ((1 - \varepsilon_t) / \varepsilon_t)$
   d. Update distribution N_{t+1} based on α_t and ε_t
4. End for
5. Output: Final hypothesis $H(x) = sum(\alpha_t * h_t(x))$ for t = 1 to T

FIGURE 9.3 AdaBoost algorithm steps.

misclassifying a randomly selected training example drawn from the distribution N_t. Based on ε_t, the algorithm computes a weight $\alpha_t = 1/2\ln((1 - \varepsilon_t)/\varepsilon_t)$ and updates the distribution N_t to N_{t+1} as follows (see Figure 9.3).

The AdaBoost algorithm employs these steps to iteratively improve the performance of the weak learners and create a strong learner that effectively combines their predictions. The algorithm then uses ε_t to compute a weight $\alpha_t = 1/2\ln(1 - \varepsilon_t/\varepsilon_t)$ and updates the distribution N_t to N_{t+1} as follows:

$$N_{t+1} = N_t / Z_t X_e - a_t, \quad \text{if } h_t(x) = c(x)$$

$$N_{t+1} = N_t / Z_t\, X_e a_t, \quad \text{if } h_t(x) \neq c(x)$$

The combined classifier $h_t(x)$ and the function $Z_t(x)$ can be computed in classical way.

$$ht(x) = \sum_{t=1}^{T} \alpha tht(x)$$

9.2.2 Gradient Boost

Gradient boosting is a strategy that stands out for its prediction speed and accuracy even for huge and complicated datasets. This method has delivered the greatest results across a range of applications including commercial machine learning systems. Errors are significant factors in major machine learning systems. The possible basic categories of errors are bias error and variance error. The gradient boost approach can be used to reduce the inaccuracy of a model's bias.

The idea of this algorithm is to build models sequentially and the subsequent models will work to rectify the errors and deviations of the previous model. Therefore, new subsequent models are built on the errors and residuals of the previous models. If the target column is continuous over a range of values, a gradient boosting regressor can be used. If the problem is of classification category, then a gradient boosting classifier can be used. Like any boosting algorithm, the goal is to minimize the loss function.

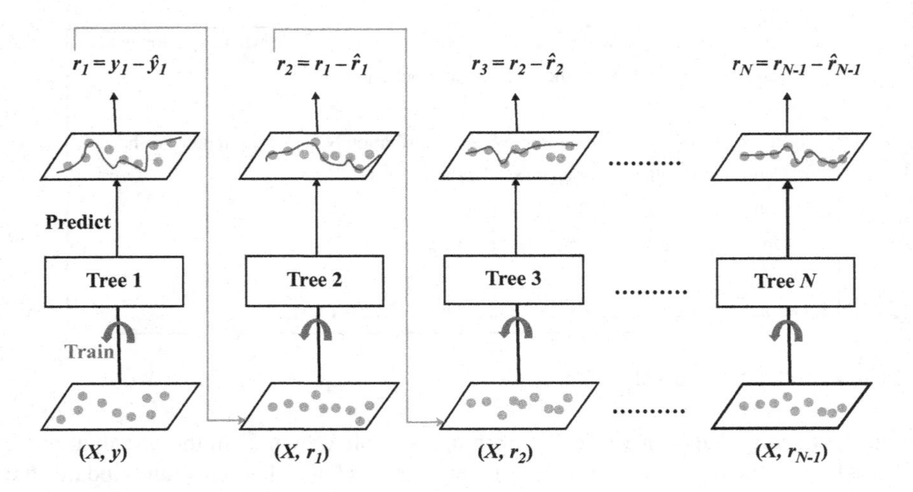

FIGURE 9.4 Model representation of repetition of residuals.

A simple loss function of gradient boost, based on the average of target column, can be defined as,

$F_0(x) = \arg\gamma\min \sum_{i=1}^{n} L(yi, \gamma)$, where L is the loss function and γ is the predicted value.

$L = 1/n \sum_{i=0}^{n} (Y_i - \gamma_i)^2$, where Y_i is the observed value and γ_i is the predicted value.

We can use trial and error to find out the minimum value for γ such that the loss function is minimum. We can also use maxima minima equation to find the minimum value for γ. Finally, the residuals can be calculated by subtracting the observed value and predicted value. If the problem is classification, a decision tree van is built from the pseudo residuals. The steps can be repeated to create a model on the residuals obtained (see Figure 9.4).

$$\text{New Predicted values} = \text{Previous predicted values} + \text{Rate of Learning} * \text{Decision Tree on Residuals}$$

Contrary to AdaBoost, each predictor is trained using the residual errors of the predecessor as labels rather than having the weights of the training instances adjusted.

9.2.3 XGBoost

The strength of a powerful method lies in its scalability, enabling efficient memory utilization and parallel and distributed computing for rapid learning. Therefore,

CERN(Conseil Européen pour la Recherche Nucléaire) identified it as the best approach for classifying signals from the LHC (Large Hadron Collider). In tackling the challenge of distinguishing rare signals from background noise in a complex physical process, CERN needed a scalable solution to analyze data generated at a rate of 3 petabytes per year. XGBoost emerged as the optimal, simplest, and most reliable choice.

XGBoost is an ensemble learning technique that recognizes the limitations of relying solely on the output of a single machine learning model. It offers a systematic approach to combine the predictive capabilities of multiple learners. The ensemble generates a consolidated output based on the predictions of various base learners, which can stem from the same or different learning algorithms. Popular ensemble learners include bagging and boosting, with decision trees being a common application historically, which will be discussed at a later stage of the chapter.

The boosting for ML and AI elaboration principles and the contrasting effects of boosting, bagging, and voting in QML are discussed in the subsequent sections. In the final section of this chapter, methods for using boosting in supervised machine learning models as well as its benefits and limitations are examined.

9.3 QUANTUM BOOST FOR ML AND AI

Artificial intelligence (AI) and machine learning (ML) have garnered a lot of attention over the past decade due to their effectiveness in real-world applications. Given the wide range of applications for machine learning, there has been considerable interest in determining which learning processes quantum computers could speed up. In this regard, there has been an increase in quantum algorithms that may offer exponentially or polynomial quantum accelerations over conventional computers for practical machine learning applications. In previous decades, empirical research on quantum machine learning (QML) has focused on developing efficient quantum algorithms with beneficial quantum complexity to handle interesting learning issues. As we are aware, the AdaBoost algorithm, is a well-known boosting algorithm with several applications in both theory and practice.

9.3.1 Enhancing the Time Complexity of the Original AdaBoost in QML

The emerging discipline of quantum machine learning (QML) has given us algorithms for many quantum and conventional learning tasks, including: (i) quantum advancements to standard methods for essentially driven machine learning operations, such as perceptron learning [4], support vector neural networks [5], and kernel-based classifiers [6, 7]; (ii) the learnability of quantum objects [8–10]; and (iii) quantum algorithms for optimization [11]. Even while these results are intriguing and demonstrate that quantum computers can significantly outperform conventional computers at solving pertinent machine learning issues, there are still a lot of challenges to be solved. Recent studies have demonstrated the dequantization of QML algorithms or the viability of conventional methods for machine learning tasks that were previously believed to reap advantages from enormous quantum acceleration [12].

The typical AdaBoost algorithm is substantially outperformed by a quantum boosting method. The quantum boosting method was used to transform a weak and inaccurate QML algorithm into a strong and accurate algorithm, achieving results tenfold faster than the conventional boosting procedures.

9.3.2 Strength of Quantum Boost Algorithm over Shortcomings of Classical AdaBoost

AdaBoost encounters the following problems when utilizing the mean estimation subroutine to reduce time complexity:

- Approximation error concerns.
- A powerful estimation of the approximates.
- Noise in the quantum learner's inputs.

The approximation-related problems are handled using a quantum method with a modified version of the classical AdaBoost standard distribution update rule. The quantum algorithm uses two quantum subroutines, the first of which, on a quantum computer, determines the mean of numbers substantially faster than the conventional AdaBoost. The generalization and training errors are successfully decreased using the quantum boosting approach.

9.3.3 Optimized Machine Learning Using Quantum Boosting

Researchers recognize that the programming style used by quantum computers in the years to come will have a profound effect on machine learning. This approach, which is a part of the wider area of inquiry termed quantum machine learning, examines the competing approach of using traditional machine learning to evaluate findings from quantum investigations, is referred to as "quantum enhanced machine learning". To understand the notion of quantum enhanced machine learning, one must appreciate machine learning's workings and the fascination of enhancing its potential.

Machine learning can be quickly introduced using data fitting. Upon importing the empirical information onto a computer, statistical software might be used to identify the parameter-dependent function $f(x)$ that best fits the data, which may result in an optimization issue. Machine learning often utilizes an optimal model function to resolve the optimization problem and predict the outcomes of assessments for novel control parameters without actually conducting a study.

Let us take the Himalayan mountain range as our optimization landscape, and imagine that we need to locate the deepest valley without a map and travelling on foot (see Figure 9.5).

Even if they reach the deepest valley, hikers will not definitely arrive at a good model, therefore, the optimization landscape that aligns with the model that is more adaptable is not particularly helpful. Even if they cannot fully forecast the seen data, optimal models that generalize from the deeper trend to hidden information are produced by an effective optimization landscape. To fully utilize the power of machine

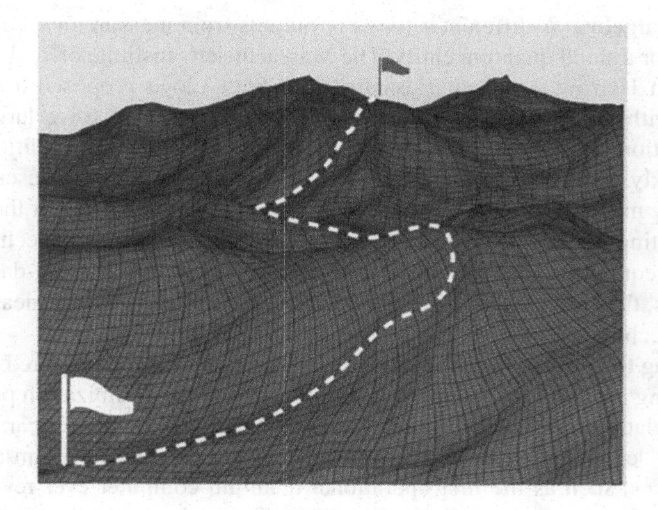

FIGURE 9.5 Optimal path proposed by machine learning.

learning, it is essential to develop strong intuition and practical expertise while formulating optimization tasks.

9.3.4 Applying the Quantum Boost

The most prevalent approach to use quantum computing to improve machine learning is to delegate the challenging optimization tasks to one of the lab-scale quantum computers currently under development or the full-scale quantum computer we anticipate using in the future. The group working on quantum data processing has created an extensive toolbox of algorithms specifically for this use. The ongoing task is to integrate, modify, and expand these technologies in order to enhance number-crunching on traditional computers.

9.3.4.1 Strategic Approaches to Quantum-enhanced Machine Learning

Quantum inquiry: The technique to search an unsorted database, such as the telephone numbers in a phone directory, with the help of a future quantum computer was proposed early in the mid-90s by a computer scientist, Lov Grover. According to this proximity, the k phone numbers having a majority of digits in commonality with the supplied number are the k "closest" entries. It is crucial for machine learning to identify the set of data points that are adjacent to a new input. For instance, the "k-nearest neighbor" strategy selects the new y-value based on the values of its contemporaries. Therefore, rewriting search issues in the conceptual framework of quantum computing and using Grover's technique to search an unorganized dataset may be the most direct method of implementing machine learning with quantum technology.

Linear algebra: *N* different settings or outputs from measurements are possible for a small quantum entity. The Massachusetts Institute of Technology's Aram Harrow, Avinatan Hassidim, and Seth Lloyd proposed a quantum algorithm that ingeniously exploits these properties to solve large linear equations, which may be achieved under highly restricted conditions quite quickly. A directed assortment of variables can be formally described for many machine learning optimization problems, with the size of the dataset dictating the number of unknown variables. Since computing technologies can require a significant amount of processing capacity for big-data workloads, they are ideally suited for the adoption of the dynamic linear modeling technique.

Locating the foundational state: An ideal bit sequence is discovered by minimizing a larger vitality effect on a third category of optimization problems. Simulation of annealing is a common numerical technique for carrying out such "combinatorial optimization". There are existing quantum-annealed devices, such as the first operational quantum computer ever revealed by the Canadian company D-Wave in 2010. These electronics have been demonstrated to identify universal minimums in issues a hundred million times speedier than a regular computer using a virtual annealing approach.

Encoding data for quantum machines: While most difficult computing issues tend to remain difficult even when quantum effects are used, small speedups might still be extremely important for today's big-data workloads. However, offshoring comes with a crucial proviso. This method requires encoding the data that determines the optimization environment into the parameters of the quantum system.

As seen in Figure 9.6, one approach to accomplish this is to portray a picture in black-and-white using a grid of spins indicating upward (white pixel) or downward (black pixel). Numerous images can be stored in an individual quantum device by employing quantum juxtaposition, in which anatomical structure is in multiple quantum phases simultaneously. Other encoding techniques tend to be more complex, but

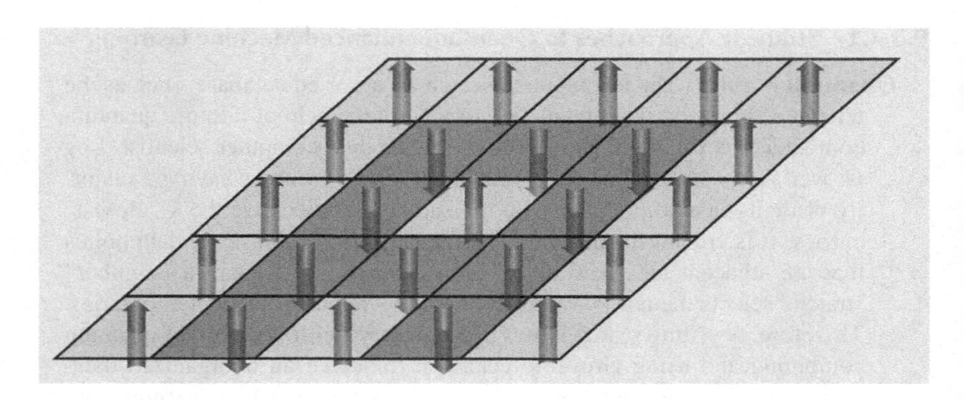

FIGURE 9.6 Representation of data encoding in the grid of spins.

they all practically entail that the quantum technique's baseline state (or, in some circumstances, the interactions) is arranged to reflect the dataset's data values. Therefore, encoded data is a key constraint of quantum calculations for machine learning, and a problem that cannot be directly solved by conventional computers [13].

9.3.5 An Upsurge of Quantum Boost with AI

The development of artificial intelligence could be considerably accelerated by future quantum computers. These computers have the ability, one day, to solve problems like decoding secret keys that are beyond the capacity of conventional computers since they store data in "fuzzy" quantum assertions that can be both zero and one at once. The vast majority of quantum methods developed to date have usually been concentrated on problems like decoding keys or perusing a list—tasks that usually address for speed but minimal intelligence. It was developed to detect commonalities in data using a quantum version of "machine learning"—a type of artificial intelligence (AI) in which programs learn from their mistakes and get better with practice. Machine learning is widely used in applications like junk mail monitors and e-commerce recommendations. It would benefit from quantum calculations to speed up machine learning processes exponentially.

A quantum computer might fool users by shortening the data and performing calculations on particular attributes that were extracted from the input and translated into qubits. Quantum machine learning takes advantage of the results of algebraic manipulations. Data can be organized, which is what speech and handwriting recognition software primarily does, or patterns can be found in the data. As a result, massive volumes of data from statistics might potentially be changed with the use of few qubits. In areas where companies like Google continue to pour a lot of resources into image identification for comparing pictures on the web or for enabling autonomous vehicles, these quantum AI techniques have the potential to drastically speed up techniques. [14].

9.3.6 Accelerating Face Recognition Speed through Boosting

The substantial margin classifier AdaBoost is effective for online learning. The original Adaboost, which makes use of all available capabilities, is contrasted with boosting along feature dimensionality as a way to tailor the AdaBoost algorithm for quick face recognition. The adoption of the latter, which is better for categorization, is ensured by identical results. Usually, the AdaBoost distinguishes between two classes. The majority-vote (MV) technique can be used to integrate all of the pairwise determinations in order to solve the multi-class detection issue (see Figure 9.7).

In contrast, when the number of persons (n) in the face database is very large, the frequency of pairwise contrasts is enormous, that is, $n(n - 1)/2$. A constrained majority voting (CMV) technique significantly lowers the number of comparisons made pairwise without sacrificing precision in recognition. Similar results can be obtained by boosting along dimensionality as opposed to employing all features in every round. As a result, evaluation and development procedures can both be greatly accelerated. Moreover, it is of no significance whether the features are weighted or not

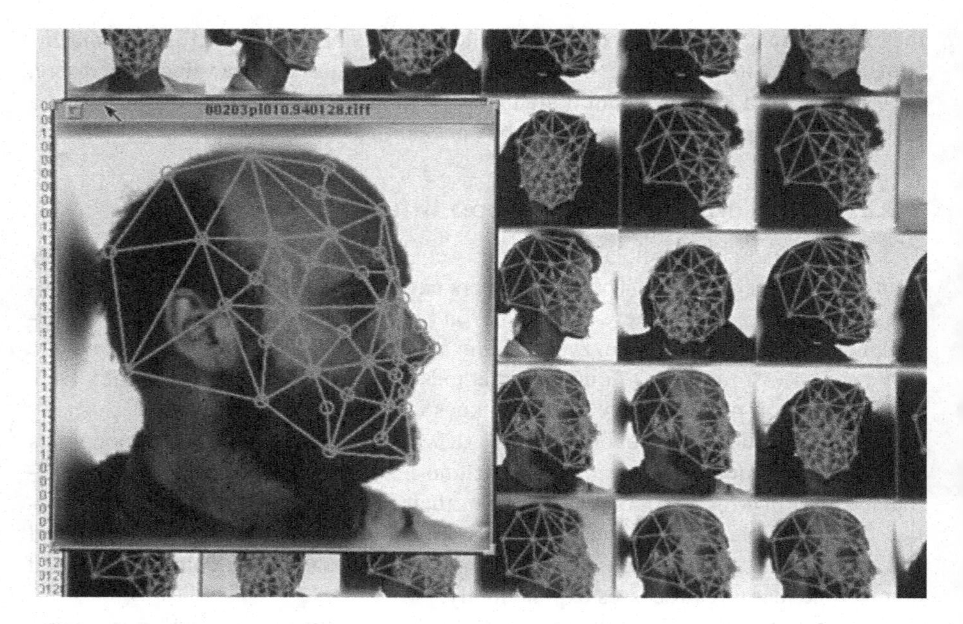

FIGURE 9.7 Fast face recognition through AdaBoost.

during the process of boosting throughout the feature dimensions. The CMV technique is quicker than the conventional voting strategy that applies all pairs, and can be utilized to speed up the multi-class facial recognition process even more, without overtly sacrificing recognition accuracy. CMVBoost.2 and CMVBoost.3 can, therefore, be employed for quick face recognition. Another finding is that over-fitting issues escalate when boosting on face data. Research is being carried out to prevent the issues associated with over-fitting.

In order to become a truly interdisciplinary venture, the emerging subject of quantum augmented machine learning must escape the limitations of quantum computing. For this, a substantial degree of contact and translation expertise is required. The statistical characteristics of observations covered by both quantum theory and machine learning may not be as dissimilar as depicted. Perhaps there is no need for to use the electronic ones and zeros as a diversion. [15].

9.4 BOOSTING vs. BAGGING vs. VOTING

Machine learning ensemble approaches like boosting, bagging, and voting are all used to enhance the performance of predictive models by aggregating the predictions of various distinct models. Each of these methods has its own methodology and traits.

The steps the boosting algorithm takes are as follows. Initialization of the training data is done, after which each data point is given an equal weight. The data is then used to train a weak learner. The weak learner is a straightforward but inaccurate paradigm. It can, however, accurately identify some of the data points.

Calculating the weak learner's error rate is the next stage. The percentage of data points that a novice learner erroneously categorized is known as the error rate. In the following iteration, the data points that the weak learner mistakenly identifies are given a higher weight. This implies that the following slow learner will concentrate on these facts and attempt to categorize them correctly.

9.4.1 Boosting

Boosting is an ensemble learning meta-algorithm primarily for reducing bias, and also variance in supervised learning. It combines a set of weak learners into a strong learner. A weak learner is a classifier that is only slightly correlated with the true classification (it can label examples better than random guessing). In contrast, a strong learner is a classifier that is arbitrarily well-correlated with the true classification.

Boosting works by training a sequence of weak learners, each of which is trained on a weighted version of the training data. The weights of the training data are adjusted after each weak learner is trained, so that the next weak learner focuses on the data points that were misclassified by the previous weak learner. This process is repeated until a desired level of accuracy is reached.

One of the most popular boosting algorithms is AdaBoost (adaptive boosting). AdaBoost works by assigning weights to each training example, such that the examples that are misclassified by the current weak learner are given more weight in the training of the next weak learner. This process is repeated until a desired level of accuracy is reached.

Boosting has been shown to be effective in a variety of machine learning tasks, including classification, regression, and ranking. It is particularly well-suited for tasks where the training data is noisy or imbalanced.

Some examples of boosting in AI:

- In **spam filtering**, boosting can be used to identify spam emails. A weak learner might be trained to identify common spam words, such as "free" or "money". The next weak learner would then be trained on the data that was misclassified by the first weak learner. This process would continue until a desired level of accuracy is reached.
- In **fraud detection**, boosting can be used to identify fraudulent transactions. A weak learner might be trained to identify transactions that are likely to be fraudulent, such as transactions that are made from new accounts or that involve large amounts of money. The next weak learner would then be trained on the data that was misclassified by the first weak learner. This process would continue until a desired level of accuracy is reached.
- In **image classification**, boosting can be used to classify images into different categories, such as cats, dogs, or cars. A weak learner might be trained to identify features that are common to a particular category of images, such as the shape of an object or the color of an object. The next weak learner would then be trained on the data that was misclassified by the first weak learner. This process would continue until a desired level of accuracy is reached.

Boosting is a powerful machine learning technique that can be used to improve the accuracy of a variety of AI tasks.

9.4.2 Bagging

Bagging, also known as bootstrap aggregating, is an ensemble learning meta-algorithm that combines a set of weak learners to create a more accurate and robust model. It works by training multiple copies of a base estimator on different subsets of the training data. Each subset is created by sampling the original training data with replacement, which means that some data points may be included in multiple subsets.

The predictions of the individual models are then averaged or combined in some other way to produce a final prediction. This averaging process helps to reduce the variance of the model, which can make it more robust to noise and outliers in the training data.

Bagging is a relatively simple and easy-to-implement ensemble learning method. It can be used with a variety of base estimators, including decision trees, random forests, and support vector machines. Bagging has been shown to be effective in a variety of machine learning tasks, including classification, regression, and clustering (see Figure 9.8).

Benefits of using bagging in machine learning:

- It can help to reduce the variance of a model, which can make it more robust with regard to noise and outliers in the training data.
- It can improve the accuracy of a model, especially when the base estimator is a weak learner.
- It can be used with a variety of base estimators, making it a versatile ensemble learning method.
- It is relatively simple and easy to implement.

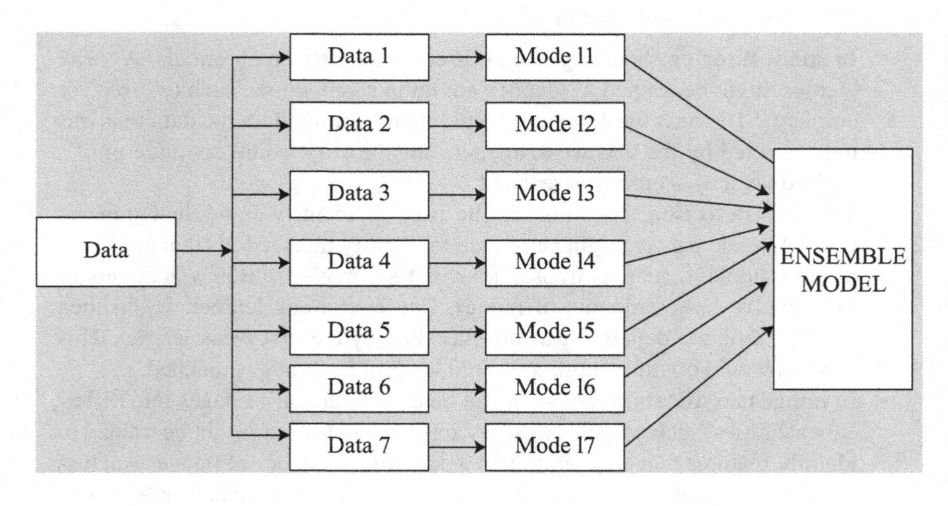

FIGURE 9.8 Bagging.

Drawbacks of using bagging in machine learning:

- It can increase the training time, as multiple models need to be trained.
- It can increase the complexity of the model, which can make it more difficult to interpret.
- It can be less effective than other ensemble learning methods, such as boosting, in some cases.

Overall, bagging is a powerful ensemble learning method that can be used to improve the accuracy and robustness of machine learning models. It is a relatively simple and easy-to-implement method that can be used with a variety of base estimators. However, it is important to note that bagging can increase the training time and complexity of the model.

9.4.3 VOTING

Voting is an ensemble learning method that combines the predictions of multiple machine learning models to make a final prediction. It is a simple and effective way to improve the accuracy of a model, especially when the individual models are diverse. There are two main types of voting—hard voting and soft voting. Hard voting simply takes the majority vote of the individual models. For example, if a model predicts class A, B, and C with equal probability and another model predicts class A and B with equal probability, then the final prediction will be class A. Soft voting takes into account the confidence of each model's prediction. For example, if a model predicts class A with a probability of 0.6, and another model predicts class B with a probability of 0.4, then the final prediction will be class A. Voting can be used with a variety of machine learning models, including decision trees, support vector machines, and logistic regression. It is especially effective when the individual models are diverse, meaning that they use different algorithms and make different types of predictions.

Benefits of using voting in machine learning:

- It can improve the accuracy of a model, especially when the individual models are diverse.
- It can be more robust to noise and outliers in the training data.
- It can be more interpretable than other ensemble learning methods, such as boosting.
- It is relatively simple and easy to implement.

Drawbacks of using voting in machine learning:

- It can increase the training time, as multiple models need to be trained.
- It can increase the complexity of the model, which can make it more difficult to train and deploy.

TABLE 9.1
Comparison of Boosting, Bagging, and Voting

Boosting	Bagging	Voting
Adopts a method of sequential ensemble learning.	Adopts a method of parallel ensemble learning.	Adopts a method for merging the outcomes of various models. Either a parallel or serial algorithm can be employed to train each individual model in the ensemble.
Uses random sampling to generate training sets.	By sampling the initial training data with replacement, uses a process known as bootstrapping by which the subsets are produced.	Can be utilized with any combination of models, including both strong and weak learners.
AdaBoost, Gradient Boosting, and XGBoost are examples of well-liked boosting algorithms.	A well-liked ensemble technique called Random Forest makes use of bagging by training many decision trees on various bootstrap sample.	The different voting procedures are: Hard voting: The final forecast is made by the class that receives the most votes. Soft voting: The class with the highest average probability is chosen after the estimated probabilities from various models are summed.
This technique is a method of increasing predictions from same types.	This technique is a straightforward method of increasing predictions from distinct types.	Multiple models' predictions are combined through voting either by picking the majority or averaging the probabilities.
Weighed according to their outcomes.	Bagging gives equal weightage.	Voting can be used to solve difficulties in classification and regression.
Boosting reduces bias but not variance.	Bagging typically reduces variance rather than bias.	The individual models may have different biases and variances whether utilizing majority voting or plurality voting in an ensemble.

- It can be less effective, in some cases, than other ensemble learning methods, such as boosting.

Overall, voting is a powerful ensemble learning method that can be used to improve the accuracy and robustness of machine learning models. It is a relatively simple and easy-to-implement method that can be used with a variety of machine learning models. However, it is important to note that voting can increase the training time and complexity of the model (see Table 9.1).

9.5 APPLYING BOOSTING IN SUPERVISED ML MODELS

Quantum support vector machines (QSVM) are a crucial instrument in the research and application of quantum kernel techniques. A boosting method can be used to create QSVM model ensembles, and performance improvements on several datasets [16] have been thoroughly assessed. Since this strategy was created using the top ensemble creation strategies that were effective in traditional machine learning, it ought to expand the efficacy frontiers of quantum models to a greater extent. A single QSVM model with precisely calibrated hyper-parameters can replicate the data in some cases, but in other cases, it is preferable to utilize a collection of QSVMs that are forced to do feature space investigation, utilizing the boosting strategy for automatic feature domain research. Here, the automatic feature space discovery method utilized in supervised quantum machine learning models is discussed.

In order to apply the notion of boosting quantum ML models in the context of adiabatic quantum computing, [3] used one level decision trees as weak classifiers in D-wave annealers. Typically, QML models include: (i) a variational quantum circuit with trainable parameters; (ii) data encoded into qubits; (iii) a classical cost function; and (iv) an optimization technique. These methods for building QML models are mathematically connected to quantum kernel techniques [17].

9.5.1 Efficiency of Initial Feature Maps

A circuit known as a feature map is used for initial state setup and eventual unified transformation with input features. In contrast to other kinds of machine learning algorithms, quantum support vector machines (QSVM) are independent from one another since the initial feature map that is chosen might produce distinct decision limits. However, given the variety of feature map alternatives, mechanization of the features mapping/model decision-making and training process is particularly desirable. This QSVM functionality makes it simple to incorporate boosting approaches. Quantum models can be made to function better, while also becoming more user-friendly, by automating the model selection and training process. Currently, framework designs are frequently developed from well-known physical models; for example, feature mapping employed an Ising model and corresponding Hamiltonian. As a result, an automated process can help individuals lacking physics expertise understand some of the unneeded complexity of the present models and aid in the discovery of new model frameworks.

Steps to be followed:

Step 1: Focus the spotlight on developing universal quantum computing systems using superconducting cubit-based gates, for example, IBM Quantum System One.

Step 2: Take into consideration the shallow circuits for kernel functions as they are more effective than standard weak classifiers, which are mostly based on decision trees.

Step 3: At each boosting step, use an automated model selection method, such as an Ising-type model, to select from a variety of topologies and investigate larger feature and model spaces.

Step 4: Employ the technique for both regression and classification tasks.

Step 5: Focus on models with improved performance as well as quantum speedup for classical processes.

TABLE 9.2
Increase in Boosting with QSVM Accuracy

Dataset	Mean	Max	Number of Ensembles with > learners (Out of 50)
XOR	4.2%	16.0%	36
Circles	2.0%	2.0%	2
Moons	7.5%	14.0%	24

9.5.2 Boosting Mechanisms for QSVMs

The boosting mechanisms for QSVMs are carried out in three steps: (i) encoding the data; (ii) ensemble structure; and (iii) obtaining the numerical simulation results (see Table 9.2).

Encoding the data: In this stage, we consider common examples of produced data, such as circles, moons, and XOR. This makes it possible to create a wide range of diverse datasets to gather metrics on model performance. According to best procedure, the data is separated into training, testing, and validation datasets. Hyper-parameter optimization is carried out at each stage of the boosting procedure using a validation dataset in a grid-like search for the best model. A testing dataset that has been completely concealed from the training dataset is used to compare multiple models.

Ensemble structure: Insights from various training models are combined in machine learning orchestra techniques to enable more accurate and superior conclusions. Decision stumps are examples of weak learners that are repeatedly reinforced in the traditional AdaBoost type of boosting [18]. By calculating, assigning, or changing their weights, it highlights samples that were previously misclassified in each subsequent cycle. The final prediction is decided by classifiers voting with a weighted majority. The SVMs are commonly employed on the quantum kernels and they are dubbed quantum support vector machines (QSVMs).

Input $X_{train}, y_{train}, X_{val}, y_{val}, y_{train,i} \in \{0,1\}, y_{val,i} \in \{0,1\}$, grid parameters for QSVM
Output $G(x)$

1: Initialize $w_i = 1, \forall i$.
2: **for** $m = 1$ to M **do**
3: Perform grid search and select the best classifier $G_m(x)$ on $(X_{train}, y_{train}, X_{val}, y_{val})$ taking into account exclusions from the grid and training weights w_i
4: Check early stopping conditions for perfect and worse than random guessing classification.
5: Exclude selected feature map from grid parameters for next iterations.
6: Compute $err_m = \frac{\sum_{i=1}^{N} w_i \cdot I(y_{train,i} \neq G_m(X_{train,i}))}{\sum_{i=1}^{N} w_i}$ *(estimator error)*
7: Compute $\alpha_m = \log((1 - err_m)/err_m)$ *(estimator weight)*
8: Set $w_i \leftarrow w_i \cdot \exp[\alpha_m \cdot I(y_{train,i} \neq G_m(X_{train,i}))]$
9: Output $G(x) = \sum_{m=1}^{M}(\alpha_m G_m(x))/\sum_{m=1}^{M}(\alpha_m)$

FIGURE 9.9 Boosted QSVM classifier algorithm.

Given that a QSVM is not a slow learner, boosting techniques are used going forward with very minor adjustments. Each subsequent iteration accentuates samples that have previously been incorrectly categorized by calculating, allocating, or changing their weights. The final prediction is computed using the weighted vote of the majority of classifiers.

Now, let us consider the support vector machines on quantum kernels given $k\left(\vec{x_i},\vec{x_j}\right)=\left|\left(U_\Phi\left(\vec{x_i}\right)|U_\Phi\left(\vec{x_j}\right)\right)\right|^2$ also known as quantum support vector machines (QSVMs). The boosting technique is modified [16] as QSVM is not a poor learner, as shown in Figure 9.9.

Grid search parameters and training and validation datasets are first provided to the algorithm. For Sklearn's support vector classifier (SVC), additional parameters include the Pauli feature map set, the Pauli rotation factor (alpha), and a regularization parameter (C). Alpha has a range of [0; 2], whereas C has a range of [1; 100]. Each example starts out with a weight of 1. To discover the optimal model, grid search is applied to a validation dataset. After choosing the best model, early halting conditions are investigated. The most recent model object is returned when any halting condition is satisfied. A weighted majority vote of the classifiers in the model can be used to generate predictions for new samples using this object. In addition, the grid search in upcoming iterations will not use the applied feature map that was used in the current iteration. When the stopping condition is met, the method changes the weights automatically and returns the final model object, which may be used to make predictions for future samples as a weighted majority vote of the model's classifiers. Based on the least amount of error on the validation sample, the ideal number of estimators is selected. As a result, in contrast to conventional boosting tactics, the ensemble approach successfully conducts a grid search for the optimal model on each iteration of the algorithm and promotes the investigation of various model designs via parameter grid constraints.

9.5.3 Efficacy of Boosted QSVM

Let us pick 50 datasets with 150 observations each for training, validation, and testing, and divide them evenly amongst those subsets. On each dataset, a boosted QSVM is used for training. For training comparison, the SVM and XGBoost approaches are used. Radial Basis Function (RBF) linear kernels are included in the parameter grid for SVM, along with the parameter grid for XgBoost [19].

Given this, the QSVM approach is used to three datasets: XOR, circles, and moons. On the XOR dataset, the effectiveness of the three models—SVM, XGBoost, and Boosted QSVM— appears to be equivalent. The boosted QSVM performs inappropriately when used with the moons dataset, but it outperforms with a median accuracy of 100% when used with the circles dataset.

According to the investigation's findings, only 31% of the Boosted QSVM models had several estimators in the ensemble. Table 9.2 displays the average and maximum ensemble size for the dataset.

The wider the ensemble appears to be, the more complex the dataset for QSVM; more than one estimator is rarely needed for circle data, whereas 3.8 estimators are required on average for moons data. The performance benefits from multiple classifiers have demonstrated that an ensemble of QSVM classifiers outperforms a single QSVM in classification accuracy. For circular data, the sample size is minimal, and even a single QSVM performs well. XOR and moons boost classification accuracy by an average of 4.2% and 7.5%, respectively.

Data scientists from a multitude of industries strive to explore the envelope in their pursuit of an exceptional machine learning model that could offer them a competitive advantage. As a result of improved feature spaces, quantum machine learning has the potential to outperform classical machine learning. The Boosted QSVM approach described here is based on the best ensemble building techniques that were successful in traditional machine learning and should improve model performance going forward. The examples shown demonstrate that boosted QSVM ensembles perform better than independent QSVMs, enabling them to match and sometimes even outperform the accuracy of non-quantum models in a number of situations.

9.6 APPLICATIONS, BENEFITS, AND CHALLENGES IN BOOSTING

9.6.1 Applications

In the context of aerial acoustic communication, a high-speed, long-range, and robust chirp spread spectrum (HRCSS) scheme for inaudible aerial acoustic communication under dynamic channels has been proposed. It innovates in the definition of a loose orthogonality condition and leverages this orthogonality to overlap multiple chirp carriers in a single time duration to form a data symbol representing multiple bits, thereby substantially promoting the data rate. Traditional classifiers have been combined with bagging and boosting methods, which are utilized in the training phase, to create the Boosting Boosting Method (GBBM). Other applications include:

- **Fraud detection:** By combining the predictions of many models, bagging, boosting, and voting can be utilized to identify fraudulent transactions.
- **Medical diagnosis:** By combining the forecasts of several medical professionals, bagging, boosting, and voting can be utilized to diagnose diseases.
- **Financial forecasting:** By aggregating the predictions of several economists, bagging, boosting, and voting can be utilized to anticipate financial markets.

9.6.2 Benefits

Ensemble learning techniques such as bagging, boosting, and voting can all be used to raise the precision of machine learning models. They provide several advantages, such as:

- **Reduced variance:** Techniques for ensemble learning can aid in lowering the variance of a model used in machine learning. This indicates that the model is more likely to generalize well to new data and is less likely to overfit the training set of data.
- **Accuracy gain:** Ensemble learning techniques can aid in raising a machine learning model's accuracy. This is due to the fact that numerous models'

predictions are integrated, which can aid in minimizing the mistakes made by each individual model.
- **Robustness:** Compared to single models, ensemble learning techniques may be more resistant to noise and outliers in training data. This allows the effects of noise and outliers to be averaged out by combining the predictions of various models.
- **Interpretability:** When compared to single models, ensemble learning techniques may be easier to understand. This is so that a more sophisticated model that is simpler to comprehend can be made by combining the predictions of different models.

It is vital to remember that ensemble learning techniques have some disadvantages. They sometimes require more processing resources than single models, which is one disadvantage. Another is that they may be more challenging to adjust than single models.

Generally speaking, ensemble learning techniques can be a useful tool for enhancing the reliability and accuracy of machine learning models. Before selecting a method for a given application, it is crucial to thoroughly weigh the advantages and disadvantages of each one.

9.6.3 Challenges

The key challenges of boosting in QML are listed below:

- **Overfitting:** If the boosting process goes on for too long or if the underlying models are complicated, overfitting may result from iteratively training models to fix the flaws of earlier models.
- **Sensitivity to noisy data:** Boosting is susceptible to noisy or outlier data points since it may give them high weights during training, which could result in predictions that are biased or inaccurate.
- **Training time:** Since each model depends on the preceding models, sequential training is necessary for boosting. When contrasted to parallel ensemble techniques like bagging, this may lengthen the training period.

9.7 CONCLUSION

In a nutshell, the developing field of quantum enhanced machine learning has to leave the confines of quantum computing and transition into a true interdisciplinary endeavor. In conclusion, the advantages of quantum boost algorithms have been described, along with methods for easing the time complexities of the original AdaBoost in QML, which makes computation more difficult. The boosting method has also been investigated for use in accelerating AI facial recognition. The differences between voting and boosting with bagging have been discussed in the final section, along with the benefits and drawbacks, plus some difficulties that may arise in practice.

REFERENCES

1. Homes A and Jokar MR, 2020, "NISQ+: Boosting quantum computing power by approximating quantum error correction", *2020 ACM/IEEE 47th Annual International Symposium on Computer Architecture (ISCA)*.
2. Arunachalam S and Maity R., 2020, "Quantum boosting", *ICML'20: Proceedings of the 37th International Conference on Machine Learning*, Article No. 36, pp. 377–387.
3. Neven H, Denchev VS, Rose G, and Macready WG, 2009, "Training a large scale classifier with the quantum adiabatic algorithm", arXiv:0912.0779.
4. Kapoor A, Weibe N, and Svore K, 2016, "Quantum perceptron models", In *Proceedings of Neural Information Processing Systems'16*, pp. 3999–4007, arxiv:1602.04799.
5. Rebentrost P, Schuld M, Wossnig L, Petruccione F, and Llyod S, 2019, "Quantum gradient descent and Newton's method for constrained polynomial optimization", *New Journal of Physics*, 1, 1–21.
6. Havlíček V, Córcoles, AD, Temme K, Harrow AW, Kandala A, Chow JM, and Gambetta JM, 2019, "Supervised learning with quantum-enhanced feature spaces", *Nature*, 567(7747):209, arXiv:1804.11326.
7. Li T, Chakrabarti S, and Wu X, 2019, "Sublinear quantum algorithms for training linear and kernek-based classifiers", In *Proceedings of the 36th International Conference on Machine Learning, ICML*, pp. 3815–3824, arXiv:1904.02276.
8. Rocchetto A, 2018, "Stabiliser states are efficiently PAC-learnable", *Quantum Information & Computation*, 18(7 & 8), 541–552.
9. Yoganathan M, 2019, "A condition under which classical simulability implifies efficient state learnability", arXiv:1907.08163.
10. Aaronson S, 2007, "The learnability of quantum states", *Proceedings of the Royal Society of London*, 463(2088), 1–12.
11. Apeldoorn JV, Gilyén A, Gribling S, and Wolf RD, 2020, "Convex optimization using quantum oracles", *Quantum*, 4(220), 312–319.
12. Chia N, Gilyén A, Li T, Lin H, Tang E, and Wang C, 2019, "Sampling-based sublinear low-rank matrix arithmetic framework for dequantizing quantum machine learning", arXiv:1910.06151.
13. Schuld M, 2017, "A quantum boost for machine learning", *Physics World*, 30(3), 28–31.
14. Powell D, 2013, "Quantum boost for artificial intelligence", *Nature*, https://doi.org/10.1038/nature.2013.13453
15. Guo G-D and Zhang H-J, 2001, "Boosting for fast face recognition", *Proceedings IEEE ICCV Workshop on Recognition, Analysis, and Tracking of Faces and Gestures in Real-Time Systems*, Vancouver, BC, Canada, pp. 96–100, https://doi.org/10.1109/RATFG.2001.938916
16. Rastunkov V, Jae-Eun P, Mishra A, Quanz B, Wood S, Codella C, Higginh H, and Broz J, 2022, "Boosting method for automated feature space discovery in supervised quantum machine learning models", *Quantum Physics*, https://doi.org/10.48550/arXiv.2205.12199
17. Schuld M, 2021, "Quantum machine learning models are kernel methods", arXiv:2101.11020.
18. Hastie T, Tibshirani R, and Friedman J, 2017, *The Elements of Statistical Learning: Data Mining, Inference and Prediction*, Springer.
19. Wade C, 2020, "*Hands-on Gradient Boosting with XGBoost and Scikit-Learn*", 1st ed., Packt Publishing.

Part IV

Quantum Evaluation Models

10 Deep Quantum Learning

Indhuja Anandan
SNS College of Technology, India

Lalith Prem Ravi
Informatics Software and Hardware, India

10.1 OPTIMIZED LEARNING BY D-WAVE

Optimized Learning by D-Wave is an innovative approach that harnesses the power of quantum annealing technology, offered by D-Wave Systems Inc., to solve complex optimization problems in machine learning and artificial intelligence. Quantum annealing is a specialized quantum computing technique that leverages quantum tunneling and thermal effects to find the optimal solutions of challenging optimization tasks. This chapter delves deeper into the principles of Optimized Learning by D-Wave, explores the mathematical foundations of quantum annealing, and examines its potential applications in machine learning.

10.1.1 Quantum Annealing and D-Wave Technology

10.1.1.1 Quantum Annealing Process

Quantum annealing is a quantum computation technique used to find the ground state of a quantum system, which corresponds to the optimal solution of an optimization problem. The process starts with the system in an initial superposition state and gradually cools it to its ground state configuration, effectively "annealing" it to the optimal solution. This process is inspired by classical simulated annealing, where a system is heated and then slowly cooled to reach a low-energy state corresponding to the optimal solution of an optimization problem. The key idea is to exploit quantum tunneling and thermal fluctuations to efficiently search for the solution space and settle into the ground state with high probability. The system evolves according to a quantum Hamiltonian that encodes the optimization problem, and the cooling is achieved by annealing the system over time. Quantum annealing is particularly well-suited to solving combinatorial optimization problems that often have vast solution spaces and are computationally challenging for classical algorithms (Figure 10.1).

DOI: 10.1201/9781003429654-14

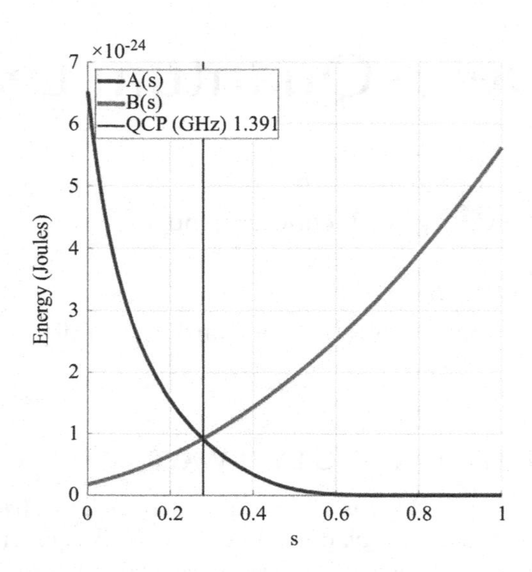

FIGURE 10.1 Quantum annealing process.

10.1.1.2 Methods of the Quantum Annealing Process

1. **Quantum Hamiltonian:** In quantum annealing, the optimization problem is mapped to a quantum Hamiltonian, denoted as $H(\tau)$, where τ is the annealing time parameter. The Hamiltonian comprises two components:
   a. **Problem Hamiltonian (H_p):** This term encodes the optimization problem and is designed such that its ground state represents the optimal solution to the problem. The problem Hamiltonian typically involves interactions between qubits that correspond to variables or constraints of the optimization problem.
   b. **Driver Hamiltonian (H_d):** The driver Hamiltonian is designed to "drive" the quantum system **during** the annealing process, making it easier to traverse the solution space. It usually involves a uniform transverse field that induces quantum fluctuations.
2. **Initial State Preparation:** At the beginning of the quantum annealing process, the quantum system is initialized in a superposition state that represents equal probabilities for all possible solutions. This initial state allows the quantum system to explore the entire solution space simultaneously.
3. **Annealing Schedule:** The annealing schedule is a time-dependent function that governs the cooling process during the quantum annealing. It determines how the quantum system evolves from the initial state to the ground state configuration. The schedule is typically controlled by a parameter $s(t)$, where t is the time during the annealing process. The annealing schedule plays a critical role in striking a balance between exploration and exploitation, enabling the system to find the global minimum efficiently.

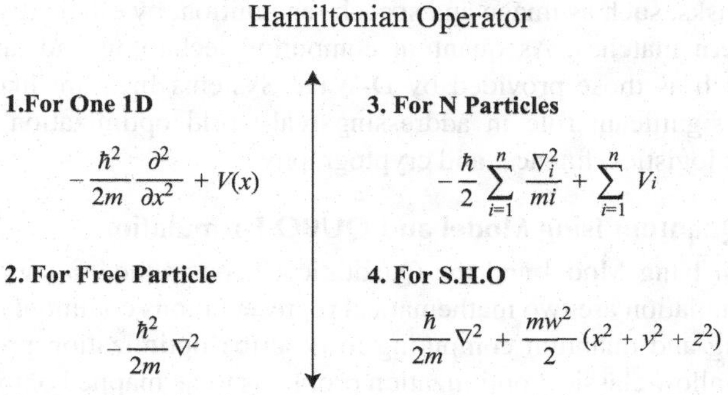

Hamiltonian Operator

1.For One 1D

$$-\frac{\hbar^2}{2m}\frac{\partial^2}{\partial x^2} + V(x)$$

2. For Free Particle

$$-\frac{\hbar^2}{2m}\nabla^2$$

3. For N Particles

$$-\frac{\hbar}{2}\sum_{i=1}^{n}\frac{\nabla_i^2}{mi} + \sum_{i=1}^{n} Vi$$

4. For S.H.O

$$-\frac{\hbar}{2m}\nabla^2 + \frac{mw^2}{2}(x^2 + y^2 + z^2)$$

FIGURE 10.2 Quantum evolution.

4. **Quantum Evolution:** As the annealing process progresses, the quantum system evolves according to the quantum Hamiltonian $H(\tau)$ based on the annealing schedule. During this evolution, the system transitions from its initial superposition state to the ground state. This evolution allows the system to explore different regions of the solution space, with a higher probability of finding the ground state or the optimal solution (Figure 10.2).

10.1.1.2.1 Practical example: Traveling Salesman Problem (TSP)

The Traveling Salesman Problem (TSP) is a classic optimization problem in which a salesman seeks the shortest route to visit a set of cities and return to the starting city, visiting each city exactly once. The goal is to find the best route for reducing the distance traveled.

Quantum annealing can be applied to the TSP by mapping the problem to an Ising model or Quadratic Unconstrained Binary Optimization (QUBO) formulation. Each city is represented by a qubit, and the interactions between qubits correspond to the distances between cities. The objective is to find the ground state of the quantum Hamiltonian, which represents the shortest route.

During the quantum annealing process, the quantum system explores different routes through quantum fluctuations and quantum tunneling. As the annealing time increases, the system gradually cools, favoring the routes with shorter distances. Eventually, the quantum system settles into the ground state, representing the optimal route that solves the TSP.

Quantum annealing is a powerful technique used to solve complex optimization problems efficiently. By exploiting quantum effects such as superposition and tunneling, quantum annealing enables the exploration of large solution spaces and provides a promising approach to tackle a wide range of combinatorial optimization problems. This is really helpful in clustering, and has shown promise in solving clustering problems, where the goal is to group data points into distinct clusters based on their similarities or distances. Feature selection optimized learning by D-Wave can be employed for feature selection in machine learning tasks. Finding the optimal subset of relevant features improves the efficiency and interpretability of models and pattern recognition, Quantum annealing offers potential applications in pattern recognition tasks, such as image and speech recognition, by efficiently searching for optimal pattern matches. As quantum computing technology advances, quantum annealers, such as those provided by D-Wave Systems Inc., are likely to play an increasingly significant role in addressing real-world optimization challenges in fields such as logistics, finance, and cryptography.

10.1.1.3 Quantum Ising Model and QUBO Formulation

The Quantum Ising Model and the Quadratic Unconstrained Binary Optimization (QUBO) formulation are two mathematical representations commonly used in quantum annealing and quantum computing to describe optimization problems. These formulations allow classical optimization problems to be mapped onto quantum systems, making them suitable for solving using quantum annealing techniques, such as those offered by D-Wave Systems Inc.

1. **Quantum Ising Model:** The Quantum Ising Model is a mathematical model derived from statistical mechanics that describes the interactions between spins on a lattice. In the context of quantum annealing, optimization problems can be mapped to the Ising model, where the goal is to find the ground state that minimizes the energy of the system (Figure 10.3).

Mathematical Example:

Consider a simple example of the Ising model on a one-dimensional chain of three spins, where each spin can be in a state of either "up" (1) or "down" (–1). Let J_{ij} represent the interaction strength between spins i and j, and h_i be the local magnetic field acting on spin i.

The given by:

$$E = \sum_i h_i \sigma_i + \sum_{(i<j)} J_{ij} \sigma_i \sigma_j,$$

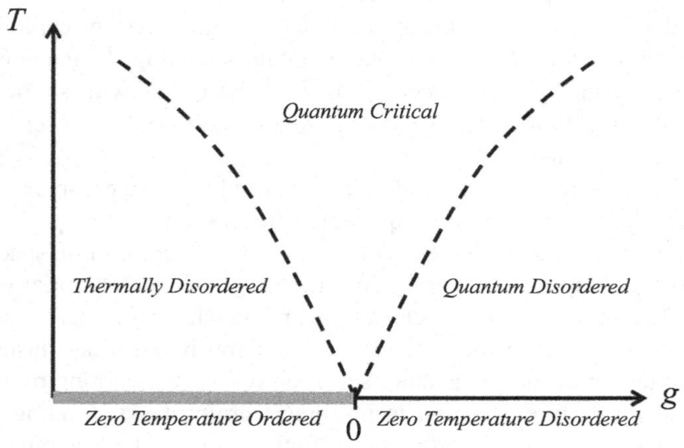

T

Quantum Critical

Thermally Disordered

Quantum Disordered

g

Zero Temperature Ordered

0

Zero Temperature Disordered

FIGURE 10.3 Quantum Ising model.

where σ_i takes values ±1 for spins up and down, respectively.

Suppose we have an optimization problem of finding the ground state that minimizes the energy. For example, let $J_{12} = -1$, $J_{23} = -2$, and $h_i = 0$ for all i. The energy expression becomes:

$$E = -\sigma_1\sigma_2 - 2\sigma_2 2\sigma_3.$$

To find the ground state that minimizes the energy, we aim to find the configuration of spins (σ_1, σ_2, σ_3) that minimizes E.

10.1.1.4 Quadratic Unconstrained Binary Optimization (QUBO) Formulation

The Quadratic Unconstrained Binary Optimization (QUBO) formulation is a mathematical representation used to convert classical optimization problems into binary quadratic optimization problems. It is particularly useful for mapping optimization problems onto quantum annealers like D-Wave quantum computers (Figure 10.4).

Mathematical Example:

Consider a simple QUBO problem with three binary variables, x_1, x_2, and x_3:

$$\text{Minimize: } C(x_1, x_2, x_3) = 2x_1 - 3x_2 + 2x_1x_2 + x_2x_3,$$

where x_i takes binary values 0 or 1.

The objective of the QUBO problem is to find the values of x_1, x_2, and x_3 that minimize the cost function $C(x_1, x_2, x_3)$.

To map this QUBO problem to the Ising model, we use the following conversions:

$$x_i = \frac{(\sigma_i + 1)}{2}$$

where σ_i takes values ±1 for spins up and down, respectively.

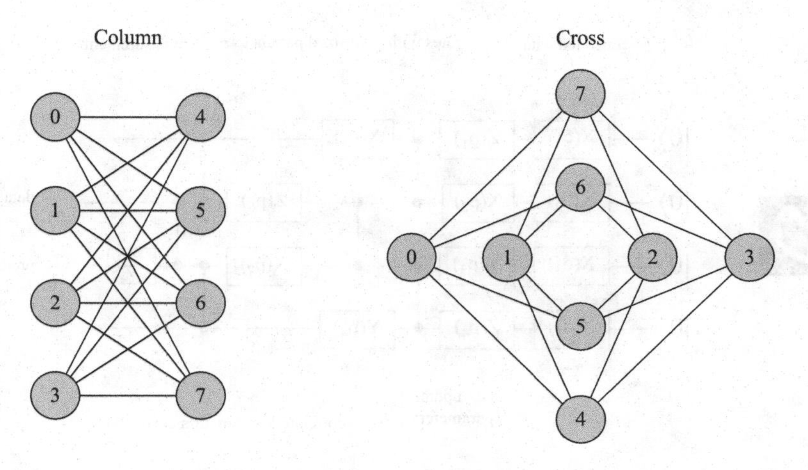

FIGURE 10.4 Quadratic unconstrained binary optimization (QUBO).

The QUBO problem is transformed into the Ising model representation:

$$E = 2(\sigma_1 - 1)(\sigma_2 - 1) - 3(\sigma_2 - 1) + 2(\sigma_1 - 1)(\sigma_2 - 1) + (\sigma_2 - 1)(\sigma_3 - 1).$$

The energy expression is now in the form of the Ising model, and the ground state of this Ising model corresponds to the optimal solution of the QUBO problem.

The Quantum Ising Model and the Quadratic Unconstrained Binary Optimization (QUBO) formulation are two powerful mathematical representations that allow classical optimization problems to be mapped onto quantum systems. By converting classical problems into quantum-compatible forms, quantum annealers, such as those provided by D-Wave Systems Inc., can be used to efficiently solve complex optimization tasks. These formulations open the door to exploring quantum advantages in solving real-world optimization problems across various domains, such as machine learning, logistics, and finance.

10.2 QUANTUM DEEP NEURAL NETWORKS

10.2.1 Introduction

Quantum deep neural networks (QDNNs) represent an innovative and interdisciplinary field that combines the principles of quantum mechanics with the power of deep learning. The idea behind QDNNs is to leverage the unique properties of quantum systems, such as superposition and entanglement, to enhance the capabilities of classical deep neural networks in solving complex problems. Thus, QDNNs hold the promise of revolutionizing various fields, including quantum computing, optimization, and machine learning, among others (Figure 10.5).

10.2.2 Classical Neural Networks vs. Quantum Neural Networks

Classical neural networks and quantum neural networks (QNNs) are two distinct paradigms for performing machine learning and data processing tasks. While both

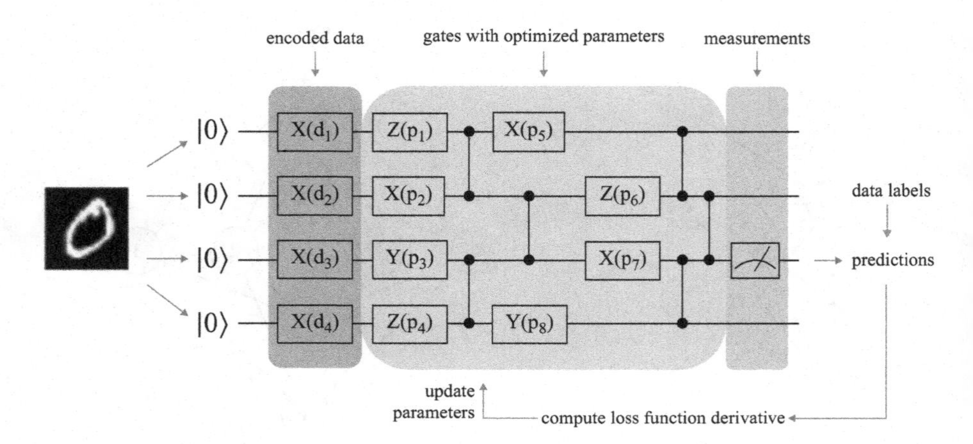

FIGURE 10.5 Quantum deep neural networks.

aim to solve complex problems, they operate on fundamentally different principles and utilize different computational models. Let's elaborate on the key differences between classical neural networks and quantum neural networks, alongside some examples.

10.2.2.1 Classical Neural Networks

1. **Operating Principle:** Classical neural networks, also known as artificial neural networks (ANNs), are based on the principles of classical computing. They consist of interconnected layers of neurons, where each neuron processes input data using weighted connections and activation functions. The network learns from data by adjusting the weights through training algorithms, such as backpropagation.
2. **Example:** A classic example of a classical neural network is a feedforward neural network used for image classification tasks. It takes an image as input and processes it through hidden layers of neurons, learning to recognize patterns and objects in the image based on the learned weights.
3. **Computational Model:** The computations in classical neural networks are deterministic, meaning they produce definite outputs for given inputs. The processing is done using classical bits, which can exist in either a 0 or 1 state, and the operations are based on classical logic.

10.2.2.2 Quantum Neural Networks

1. **Operating Principle:** Quantum neural networks, on the other hand, utilize the principles of quantum mechanics for computation. In quantum neural networks, quantum neurons represent quantum states, which can exist in multiple states simultaneously due to superposition, and they can be entangled to exhibit strong correlations.
2. **Examples:** A QNN example is a quantum circuit designed to solve optimization problems using quantum annealing. The quantum neurons are initialized in quantum superposition states, and quantum gates are applied to process the quantum data and find the optimal solution.
3. **Computational Model:** Quantum neural networks exploit the inherent parallelism and entanglement of quantum systems to perform computations more efficiently in some cases, especially for certain optimization tasks. The computations are probabilistic, and the output probabilities are obtained through measurements on the quantum states.

10.2.2.3 Differences and Advantages

1. **Superposition and Entanglement:** One key difference between classical and quantum neural networks lies in the superposition and entanglement properties of quantum systems. Quantum neural networks can explore multiple possibilities simultaneously due to superposition, potentially providing computational advantages for certain problems.
2. **Data Encoding:** In classical neural networks, data is represented using classical bits (0 or 1), while in quantum neural networks, data encoding

techniques map classical data into quantum states, taking advantage of quantum parallelism.

3. **Computational Complexity:** Quantum neural networks have the potential to offer significant speedup for certain tasks, such as quantum simulation and optimization, by exploiting quantum parallelism and the quantum nature of the problem.

Classical neural networks and QNNs are two distinct computational paradigms, each with its strengths and limitations. Classical neural networks excel in many classical machine learning tasks and have been successfully applied in various domains. Quantum neural networks, on the other hand, represent a promising avenue that explores the potential of quantum computing in solving computationally complex problems. As quantum computing technology advances, QNNs hold the potential to revolutionize various fields and drive innovations in artificial intelligence, optimization, cryptography, and more.

10.2.3 Quantum Neurons in Deep Neural Networks

Quantum neurons are the fundamental building blocks of quantum deep neural networks (QDNNs). They replace classical neurons used in conventional deep neural networks with quantum states and quantum operations. Quantum neurons leverage the principles of quantum superposition and entanglement to process information in parallel and potentially offer computational advantages over classical counterparts (Figure 10.6).

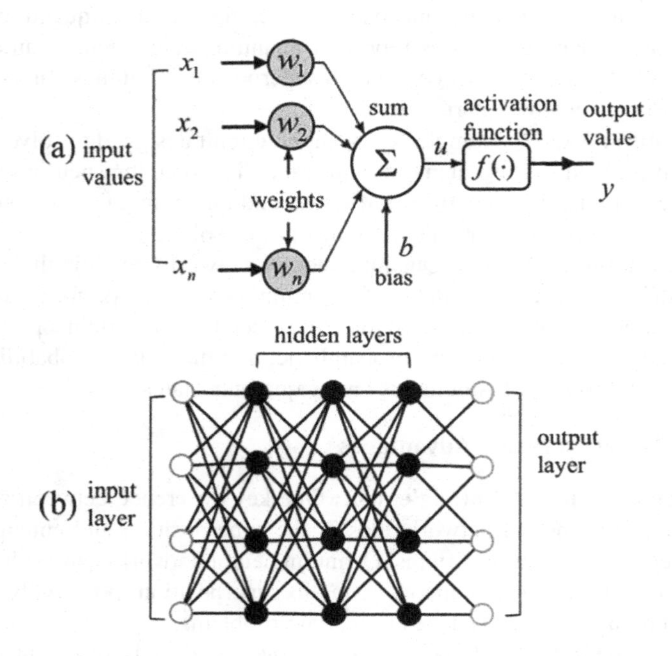

FIGURE 10.6 Quantum neurons in deep neural networks.

10.2.3.1 Quantum State Representation of Quantum Neurons

In quantum deep neural networks, a quantum neuron represents its state using qubits. The state of a quantum neuron can be described as a linear superposition of different quantum states, with complex probability amplitudes representing the coefficients of the states.

Mathematical Example:

Consider a quantum neuron with two qubits, denoted as

$$|\psi\rangle = \alpha|0\rangle + \beta|1\rangle \text{ and } |\varphi\rangle = \gamma|0\rangle + \delta|1\rangle.$$

The combined state of the quantum neuron can be expressed as

$$|\Psi\rangle = \alpha|0\rangle \otimes \gamma|0\rangle + \alpha|0\rangle \otimes \delta|1\rangle + \beta|1\rangle \otimes \gamma|0\rangle + \beta|1\rangle \otimes \delta|1\rangle.$$

10.2.3.2 Quantum Operations and Gates on Quantum Neurons

Quantum neurons undergo transformations using quantum gates and operations, similar to how classical neurons are activated by activation functions in classical deep neural networks. Quantum gates, such as the Hadamard gate, Pauli-X gate, Pauli-Y gate, Pauli-Z gate, and controlled-NOT (CNOT) gate, are applied to manipulate the quantum states.

Mathematical Example:

Let's apply a Hadamard gate (*H*) to the first qubit of the quantum neuron's state $|\Psi\rangle$. The transformation would be:

$$H(|0\rangle) = (|0\rangle + |1\rangle)/\sqrt{2},$$

$$H(|1\rangle) = (|0\rangle - |1\rangle)/\sqrt{2},$$

Applying the Hadamard gate to the first qubit, the new state of the quantum neuron becomes:

$$|\Psi\rangle' = (\alpha|0\rangle + \beta|1\rangle) \otimes ((|0\rangle + |1\rangle)/\sqrt{2}) + (\alpha|0\rangle + \beta|1\rangle) \otimes ((|0\rangle - |1\rangle)/\sqrt{2})$$
$$+ (\gamma|0\rangle + \delta|1\rangle) \otimes ((|0\rangle + |1\rangle)/\sqrt{2}) + (\gamma|0\rangle + \delta|1\rangle) \otimes ((|0\rangle - |1\rangle)/\sqrt{2}).$$

10.2.3.3 Quantum Measurement and Probabilistic Outputs

In quantum computation, measurement is a crucial step to extract classical information from quantum states. Quantum neurons can be measured to obtain classical probabilistic outputs based on the quantum state's probability amplitudes.

Mathematical Example:

After applying the Hadamard gate, the state $|\Psi\rangle'$ of the quantum neuron is:

$$|\Psi\rangle' = (1/\sqrt{2})(\alpha|00\rangle + \alpha|01\rangle + \beta|10\rangle + \beta|11\rangle + \gamma|00\rangle - \gamma|01\rangle + \delta|10\rangle - \delta|11\rangle).$$

When the quantum neuron is measured, the outcome would be one of the possible states with corresponding probabilities $|\alpha|^2$, $|\beta|^2$, $|\gamma|^2$, and $|\delta|^2$. The measurement result gives a probabilistic output for the quantum neuron.

Quantum neurons in deep neural networks are an essential component of quantum deep learning, harnessing the power of quantum superposition and entanglement to process information efficiently. By leveraging quantum properties, quantum neurons have the potential to provide computational advantages over classical neurons in solving specific problems. Quantum deep neural networks are an exciting area of research that bridges quantum computing with deep learning, promising advancements in various fields, including optimization, pattern recognition, and machine learning.

10.2.4 Quantum Operations and Quantum Gates

In quantum computing, quantum operations are fundamental transformations applied to quantum states to manipulate and process information. These operations are implemented using quantum gates, which are analogous to logic gates in classical computing. Quantum gates act on qubits, the basic units of quantum information, to perform specific transformations and computations.

10.2.4.1 Quantum Operations

Quantum operations are unitary transformations that preserve the norm of quantum states, ensuring that the total probability of finding the system in any state remains 1. These operations are reversible, meaning that they can be undone by applying the inverse operation. Some common quantum operations include (Figure 10.7):

1. **Hadamard Gate (*H*):** The Hadamard gate is a fundamental quantum gate that creates superposition by evenly distributing the probability amplitudes of $|0\rangle$ and $|1\rangle$ states. It is represented as:

$$H = \frac{1}{\sqrt{2}} * \begin{bmatrix} 1 & 1 \\ 1 & -1 \end{bmatrix}$$

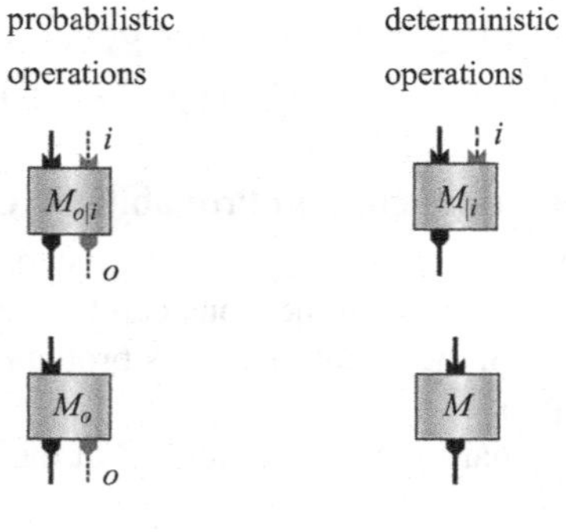

FIGURE 10.7 Quantum operations.

When applied to a single qubit, the Hadamard gate transforms the states as follows:

$$H|0\rangle = 1/\sqrt{2}(|0\rangle + |1\rangle),$$

$$H|1\rangle = 1/\sqrt{2}(|0\rangle - |1\rangle).$$

2. **Pauli Gates (*X*, *Y*, *Z*):** The Pauli gates are single-qubit gates that correspond to rotations around the *X*, *Y*, and *Z* axes of the Bloch sphere, respectively. They are represented as:

$$X = \begin{bmatrix} 0 & 1 \\ 1 & 0 \end{bmatrix},$$

$$Y = \begin{bmatrix} 0 & -i \\ i & 0 \end{bmatrix},$$

$$Z = \begin{bmatrix} 1 & 0 \\ 0 & -1 \end{bmatrix},$$

When applied to a single qubit, the Pauli gates transform the states as follows:

$$X|0\rangle = |1\rangle,$$

$$X|1\rangle = |0\rangle,$$

$$Y|0\rangle = i|1\rangle,$$

$$Y|1\rangle = -i|0\rangle,$$

$$Z|0\rangle = |0\rangle,$$

$$Z|1\rangle = -|1\rangle.$$

3. **Controlled-NOT (CNOT) Gate:** The CNOT gate is a two-qubit gate that applies the Pauli-X gate to the target qubit (second qubit) if and only if the control qubit (first qubit) is in the state $|1\rangle$. It is represented as:

$$\text{CNOT} = \begin{bmatrix} 1 & 0 & 0 & 0 \\ 0 & 1 & 0 & 0 \\ 0 & 0 & 0 & 1 \\ 0 & 0 & 1 & 0 \end{bmatrix}.$$

When applied to two qubits in the state $|00\rangle$, $|01\rangle$, $|10\rangle$, and $|11\rangle$, the CNOT gate performs the following transformations:

$$\text{CNOT}|00\rangle = |00\rangle,$$

$$\text{CNOT}|01\rangle = |01\rangle,$$

$$\text{CNOT}|10\rangle = |11\rangle,$$

$$\text{CNOT}|11\rangle = |10\rangle.$$

10.2.5 Quantum Gates in Quantum Circuits

Quantum circuits are constructed using a combination of quantum gates, enabling complex quantum computations. Quantum gates are applied sequentially to perform specific operations on qubits. Quantum circuits are designed to solve quantum algorithms, simulate quantum systems, and perform various quantum tasks (Figure 10.8).

Example Quantum Circuit:

Consider a simple quantum circuit consisting of two qubits, represented as $|\psi\rangle = \alpha|00\rangle + \beta|01\rangle + \gamma|10\rangle + \delta|11\rangle$. To apply the Hadamard gate (H) to the first qubit, the quantum circuit is as follows:

Apply Hadamard gate (H) to the first qubit:

$$H \otimes I \,(\text{identity gate})\,\text{applied to}\,|\psi\rangle.$$

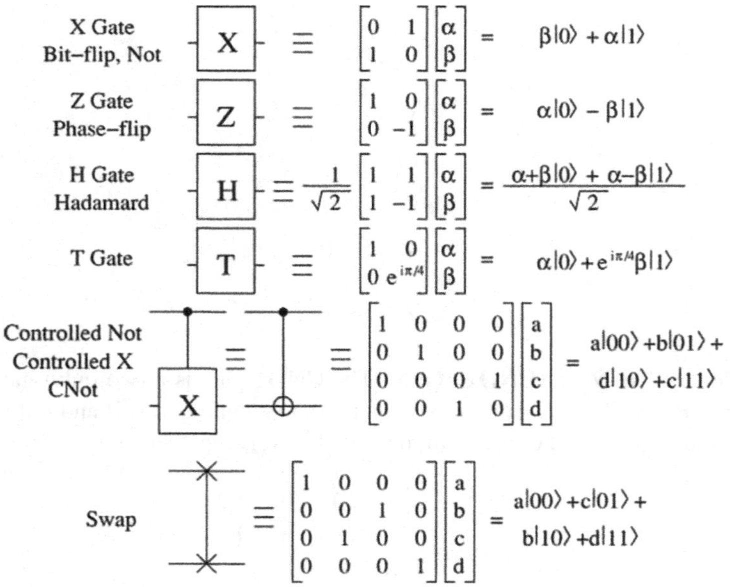

FIGURE 10.8 Quantum gates in quantum circuits.

The resulting quantum state would be:

$$H|\psi\rangle = (H \otimes I)(\alpha|00\rangle + \beta|01\rangle + \gamma|10\rangle + \delta|11\rangle) = (\alpha + \beta)|00\rangle + (\gamma + \delta)|10\rangle.$$

In this example, the Hadamard gate creates a superposition of $|00\rangle$ and $|10\rangle$ states, demonstrating the power of quantum gates to manipulate quantum states and perform quantum operations.

Quantum operations and quantum gates are fundamental building blocks of quantum computing. They enable the manipulation and processing of quantum information and play a critical role in quantum algorithms and quantum circuits. Quantum gates, such as the Hadamard gate, Pauli gates, and CNOT gate, perform specific transformations on qubits, allowing quantum systems to leverage unique quantum properties, such as superposition and entanglement, for solving complex problems and outperforming classical counterparts in specific tasks.

10.2.5.1 Quantum Circuit Layers

Quantum circuit layers are essential components of quantum algorithms and quantum computations. A quantum circuit is composed of a sequence of quantum gates applied to qubits, and these gates are typically grouped into layers to facilitate the execution of specific operations on quantum states. Each layer represents a specific transformation of the quantum state and can be customized based on the problem being solved or the quantum algorithm being executed (Figure 10.9).

1. **Structure of Quantum Circuit Layers:**
   - ***Layered quantum gates***: In a quantum circuit, quantum gates are organized into layers, where each layer consists of multiple gates applied sequentially. Gates within the same layer are typically applied in parallel to qubits, allowing for quantum parallelism and efficient computation.
   - ***Depth and width of quantum circuits***: The depth of a quantum circuit refers to the number of layers, and the width refers to the number of

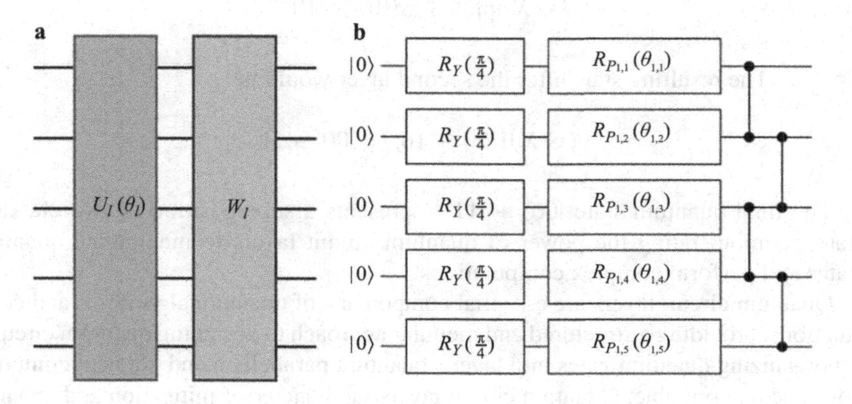

FIGURE 10.9 Quantum circuit layers.

qubits in the circuit. The depth of the circuit determines the number of sequential steps required to perform a computation, while the width defines the number of qubits involved in each step.

2. **Advantages of Quantum Circuit Layers:**
   - ***Modular design*:** Quantum circuit layers allow for a modular design, where different layers can be designed independently to perform specific operations on qubits. This modular structure makes it easier to design and optimize quantum circuits for specific tasks and quantum algorithms.
   - ***Error mitigation*:** Layered quantum circuits can be designed with error mitigation techniques in mind. By grouping gates into layers, it becomes possible to identify and correct errors that may arise during the quantum computation, enhancing the robustness and reliability of the quantum circuit.
3. **Mathematical Examples of Quantum Circuit Layers:** Let's consider a simple example of a quantum circuit with two qubits and two layers of gates:
   - ***Quantum State Initialization*:** Start with the initial state $|\psi\rangle = |00\rangle$, where both qubits are in the state $|0\rangle$.
   - ***First Layer*:** Hadamard Gate (H) and Controlled-NOT (CNOT) Gate: In the first layer, apply the Hadamard gate (H) to the first qubit and the CNOT gate to entangle the qubits:

   $$H \otimes I \,(\text{identity gate})\,\text{applied to}\, |\psi\rangle.$$

   The resulting state after the first layer would be:

   $$(H \otimes I)|\psi\rangle = (H|0\rangle) \otimes |0\rangle = (|0\rangle + |1\rangle) \otimes |0\rangle = |00\rangle + |10\rangle.$$

   - ***Second Layer*:** Pauli-X Gate and Controlled-Z Gate: In the second layer, apply the Pauli-X gate (X) to the second qubit and the Controlled-Z gate to entangle the qubits further:

   $$I \otimes X \,\text{applied to}\, (|00\rangle + |10\rangle).$$

   The resulting state after the second layer would be:

   $$(I \otimes X)(|00\rangle + |10\rangle) = |00\rangle + |11\rangle.$$

The final quantum state $|00\rangle + |11\rangle$ represents a superposition of two classical states, demonstrating the power of quantum circuit layers to manipulate quantum states and perform complex computations.

Quantum circuit layers are essential components of quantum algorithms and computations, providing a structured and modular approach to designing quantum circuits. By organizing quantum gates into layers, quantum parallelism and efficient computations become possible. Quantum circuit layers facilitate error mitigation and enhance the reliability of quantum computations. Mathematical examples demonstrate how

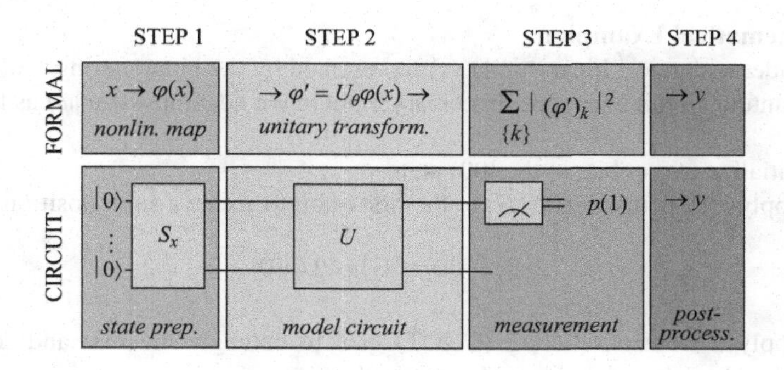

FIGURE 10.10 Quantum data encoding.

quantum circuit layers can manipulate quantum states, leading to superposition and entanglement, which are key features of quantum computing. The ability to design and optimize quantum circuit layers is crucial for advancing quantum algorithms and harnessing the potential of quantum computing for solving real-world problems.

10.2.6 Quantum Data Encoding

Quantum data encoding is a crucial step in quantum computing and quantum algorithms, where classical data is represented as quantum states. Proper data encoding is essential for efficiently processing classical information using quantum operations and algorithms. Various techniques, such as amplitude encoding and quantum circuit encoding, are employed for quantum data encoding (Figure 10.10).

10.2.6.1 Amplitude Encoding

Amplitude encoding is a popular technique for encoding classical data into quantum states by manipulating the probability amplitudes of quantum states. In amplitude encoding, classical data is mapped to the amplitudes of quantum states, which can then be processed using quantum gates and operations.

Mathematical Example:

Consider a simple example of encoding a classical binary input (x) into a single qubit quantum state. For $x = 0$, the quantum state would be $|\psi\rangle = |0\rangle$, and for $x = 1$, the quantum state would be $|\psi\rangle = |1\rangle$.

The probability amplitudes of the quantum states are $\alpha = 1$ for $|0\rangle$ and $\beta = 1$ for $|1\rangle$. The quantum state can be represented as:

$$|\psi\rangle = \alpha|0\rangle + \beta|1\rangle.$$

The classical binary input (x) is directly encoded into the probability amplitudes of the quantum state, allowing quantum operations to process the classical information.

10.2.6.2 Quantum Circuit Encoding

Quantum circuit encoding is another technique used to encode classical data into quantum states. In this method, quantum circuits are designed to perform specific unitary transformations that encode classical data into quantum states.

Mathematical Example:

Consider a classical binary input (x) represented by the binary string '10'. We can use a quantum circuit to encode this binary input into a quantum state $|\psi\rangle$ as follows:

1. Initialize two qubits in the $|00\rangle$ state.
2. Apply a Hadamard gate (H) to the first qubit to create a superposition state:

$$|\psi_1\rangle = (|0\rangle + |1\rangle)/\sqrt{2} \otimes |0\rangle.$$

3. Apply a Controlled-NOT (CNOT) gate to entangle the first and second qubits based on the classical input '10':

$$|\psi_2\rangle = (|0\rangle + |1\rangle)/\sqrt{2} \otimes |0\rangle \rightarrow (|0\rangle + |1\rangle)/\sqrt{2} \otimes |1\rangle.$$

The resulting quantum state $|\psi_2\rangle$ now encodes the classical binary input '10' into the probability amplitudes of the quantum state.

Proofs for quantum data encoding typically rely on the principles of quantum mechanics and the mathematical framework of quantum states and quantum gates. The validity of quantum data encoding techniques is rooted in the fundamental principles of quantum computing, including the superposition and entanglement of quantum states.

Quantum data encoding is a crucial step in quantum computing, enabling the representation and manipulation of classical information using quantum states and operations. Amplitude encoding and quantum circuit encoding are two widely used techniques for quantum data encoding. These techniques allow classical data to be efficiently processed using quantum algorithms, paving the way for the development of quantum-enhanced solutions in various fields, including cryptography, optimization, and machine learning. The validity of quantum data encoding techniques is established based on the mathematical foundations of quantum mechanics and the principles of quantum computing.

10.2.7 Applications of Quantum Deep Neural Networks

Quantum deep neural networks (QDNNs) hold the potential to revolutionize various fields by combining the power of quantum computing with the flexibility and versatility of deep learning. These quantum-enhanced neural networks offer unique advantages over classical counterparts in specific applications. Let's explore some of the potential applications of QDNNs:

1. ***Quantum chemistry***: Quantum chemistry involves simulating the behavior of molecules and chemical reactions and QDNNs can be applied to efficiently model quantum systems and simulate molecular properties. Quantum variational circuits can be used to optimize molecular states and calculate molecular energies, making it possible to address complex chemical problems more efficiently than classical methods.
2. ***Optimization and combinatorial problems***: Quantum deep neural networks have the potential to excel in solving optimization and combinatorial

problems. They can tackle tasks like traveling salesman problems, vehicle routing, and resource allocation. Quantum annealing techniques, combined with QDNNs, offer the potential to find optimal solutions faster and more efficiently than classical algorithms, especially for large-scale and complex optimization problems.

3. ***Quantum machine learning*****:** Quantum deep neural networks can enhance classical machine learning tasks by leveraging quantum parallelism and entanglement. Quantum-enhanced machine learning algorithms promise faster training and better generalization. Quantum data encoding and quantum feature maps enable the representation of classical data in quantum states, facilitating quantum machine learning tasks.
4. ***Quantum image and signal processing*****:** Quantum deep neural networks can be applied to image and signal processing tasks, such as image recognition, denoising, and super-resolution. Quantum circuits can be designed to process quantum representations of images and signals, potentially providing advantages over classical approaches in specific image processing tasks.
5. ***Quantum cryptography*****:** Quantum cryptography aims to ensure secure communication by leveraging quantum properties like entanglement and the no-cloning theorem. Quantum deep neural networks can be utilized to improve quantum key distribution protocols and enhance the security of quantum communication.
6. ***Quantum generative models*****:** Quantum deep neural networks can be applied to generate samples from complex quantum distributions. Quantum generative models hold the promise of generating quantum states that are challenging for classical methods to represent, contributing to advancements in quantum simulation and quantum sampling tasks.
7. ***Quantum reinforcement learning*****:** Quantum deep neural networks can be combined with reinforcement learning algorithms to optimize decision-making processes in quantum environments. Quantum reinforcement learning opens up new possibilities for solving quantum control problems and quantum optimization tasks.

Quantum deep neural networks have diverse applications across various fields, ranging from quantum chemistry and optimization to quantum machine learning and cryptography. These applications harness the unique quantum properties of superposition, entanglement, and quantum parallelism to outperform classical approaches in specific tasks. As quantum computing technology continues to advance, quantum deep neural networks hold the potential to drive innovations in both quantum computing and classical artificial intelligence, paving the way for solving complex real-world problems more efficiently and accurately.

10.3 QUANTUM CONVOLUTIONAL NEURAL NETWORKS

Quantum convolutional neural networks (QCNNs) are a specialized class of quantum machine learning models that integrate the principles of convolutional neural networks (CNNs) with quantum computing. Further, QCNNs aim to leverage the

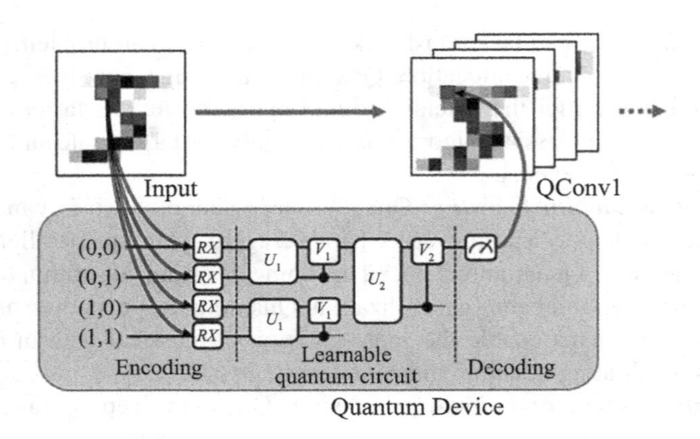

FIGURE 10.11 Quantum convolutional neural networks.

advantages of both classical deep learning and quantum computation to tackle complex problems, particularly in the realm of image and pattern recognition. This explanation dives into the concepts, architecture, and potential applications of quantum convolutional neural networks (Figure 10.11).

10.3.1 Concepts of Quantum Convolutional Neural Networks

Convolutional neural networks (CNNs) are a powerful class of deep learning models widely used for computer vision tasks, such as image classification, object detection, and image segmentation. Convolutional neural networks are specifically designed to handle grid-like data, like images, and are capable of automatically learning hierarchical features from the input data. The key components of CNNs are convolutional layers, pooling layers, and fully connected layers.

10.3.1.1 Convolutional Layers

Convolutional layers are the building blocks of CNNs. They consist of a set of learnable filters (also called kernels) that slide over the input image to extract local features. The convolution operation involves element-wise multiplication between the filter and a local region of the input image, followed by summation. This process generates feature maps that represent the presence of specific patterns in the image.

Mathematical Example:

Suppose we have a 3 × 3 filter defined as:

$$\text{Filter} = \begin{bmatrix} 1 & 0 & -1 \\ 1 & 0 & -1 \\ 1 & 0 & -1 \end{bmatrix},$$

Let's apply this filter to a 5 × 5 grayscale image:

$$\text{Image} = \begin{bmatrix} 120 & 130 & 140 & 150 & 160 \\ 110 & 120 & 130 & 140 & 150 \\ 100 & 110 & 120 & 130 & 140 \\ 90 & 100 & 110 & 120 & 130 \\ 80 & 90 & 100 & 110 & 120 \end{bmatrix},$$

The convolution operation is performed by sliding the filter over the image, performing element-wise multiplication, and summing the results to generate a feature map.

The feature map for the convolutional operation will be:

$$\text{Feature Map} = \begin{bmatrix} -40 & -30 & -20 \\ -30 & -20 & -10 \\ -20 & -10 & 0 \end{bmatrix},$$

10.3.1.2 Pooling Layers

Pooling layers help reduce the spatial dimensions of the feature maps obtained from convolutional layers. Pooling is typically done to downsample the data and make the network more computationally efficient while retaining the most important features. Max pooling is a common pooling technique, where the maximum value in a local region is selected to represent that region in the downsampled feature map (Figure 10.12).

Mathematical Example:

Suppose we have a 2 × 2 max pooling operation applied to the above feature map:

$$\text{Feature Map} = \begin{bmatrix} -40 & -30 \\ -30 & -20 \end{bmatrix},$$

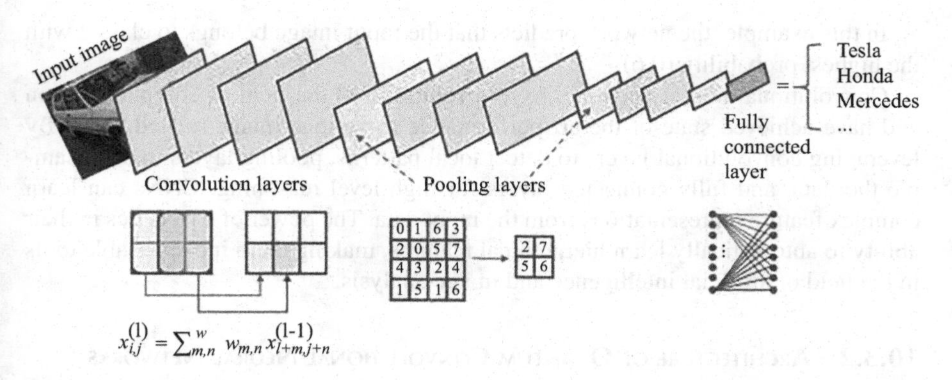

FIGURE 10.12 Pooling layers.

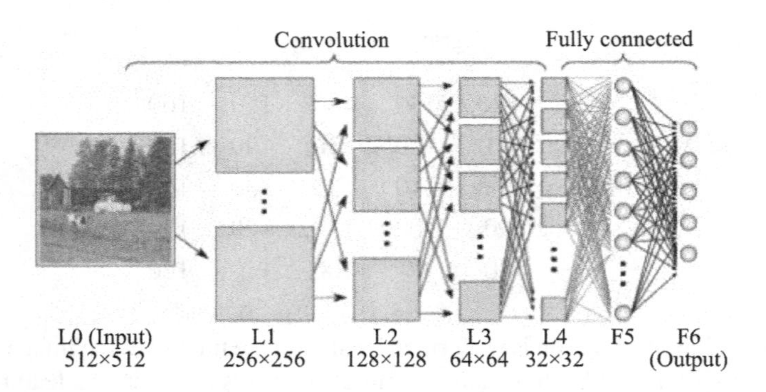

FIGURE 10.13 Fully connected layers.

After applying max pooling, the downsampled feature map will be:

$$\text{Pooled Feature Map} = \begin{bmatrix} -30 \\ -20 \end{bmatrix},$$

10.3.1.3 Fully Connected Layers

After several convolutional and pooling layers, the feature maps are flattened into a one-dimensional vector and fed into fully connected layers, which perform high-level reasoning and decision-making. The past layer's neuron is connected to the present neuron. The output of the fully connected layers provides the final predictions for the given input data (Figure 10.13).

Mathematical Example:

Suppose the final fully connected layer takes a flattened feature vector of size 100 and produces a one-hot encoded vector representing class probabilities for a classification task.

$$\text{Output} = [0.1, 0.6, 0.3, 0.0, 0.0].$$

In this example, the network predicts that the input image belongs to class 2 with the highest probability (0.6).

Convolutional neural networks have revolutionized the field of computer vision and have achieved state-of-the-art performance in various image-related tasks. By leveraging convolutional layers to detect local patterns, pooling layers to downsample the data, and fully connected layers for high-level reasoning, CNNs can learn complex feature representations from the input data. The power of CNNs lies in their ability to automatically learn hierarchical features, making them indispensable tools in the field of artificial intelligence and image analysis.

10.3.2 Architecture of Quantum Convolutional Neural Networks

The architecture of quantum convolutional neural networks (QCNNs) is a specialized design that combines the principles of classical convolutional neural networks

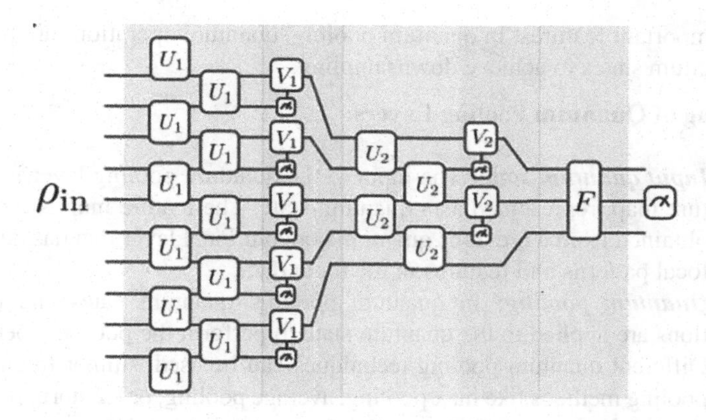

FIGURE 10.14 Architecture of Quantum Convolutional Neural Networks.

(CNNs) with quantum computing techniques to process and analyze image data in a quantum-enhanced manner. Thus, QCNNs aim to leverage the advantages of both classical deep learning and quantum computation to tackle complex computer vision tasks more efficiently. Let's explore the key components and the working of the QCNN architecture in detail (Figure 10.14).

10.3.2.1 Quantum Convolutional Layers

The quantum convolutional layer is the core building block of QCNNs. It replaces the classical convolutional layer used in traditional CNNs with quantum operations and quantum states. In classical CNNs, convolutional layers use learnable filters to slide over the input image and extract local features. In QCNNs, quantum circuits replace these learnable filters, and quantum gates perform the convolution operation.

1. ***Input quantum state*:** The input image data is encoded into quantum states using quantum data encoding techniques like amplitude encoding or quantum circuit encoding. For example, an image pixel value may be mapped to the probability amplitude of a qubit.
2. ***Quantum convolution*:** The quantum gates in the quantum circuit perform the convolution operation on the input quantum state. These quantum gates manipulate the quantum state to detect local patterns and features in the image.
3. ***Feature map*:** The result of the quantum convolution operation is a feature map, which represents the presence of specific patterns or features in the input image. This feature map is also represented as a quantum state.

10.3.2.2 Quantum Pooling Layers

Quantum pooling layers are an essential component of QCNNs. Similar to classical CNNs, quantum pooling layers downsample the feature maps obtained from quantum convolutional layers. Quantum pooling helps reduce the spatial dimensions of the data, making it more manageable and computationally efficient while retaining the most important features. In quantum pooling, quantum operations are performed on the quantum states to achieve downsampling.

Working of Quantum Pooling Layers:

1. ***Input quantum state*:** The input to the quantum pooling layer is a feature map, represented as a quantum state. The feature map is typically obtained from a previous quantum convolutional layer that has detected local patterns and features in the input data.
2. ***Quantum pooling*:** In quantum pooling, quantum gates and operations are applied to the quantum state to perform the pooling operation. Different quantum pooling techniques can be used, similar to classical pooling methods like max pooling, average pooling, or L2-norm pooling.
3. ***Pooled feature map*:** The result of the quantum pooling operation is a downsampled feature map, also represented as a quantum state. This downsampled feature map retains the essential features of the input data while reducing its spatial dimensions.

Example—Max Quantum Pooling:

Let's consider an example of max quantum pooling using a 2×2 pooling window. We have an input feature map as follows:

Input Feature Map (Quantum State):

$$\begin{bmatrix} 0.7, & 0.2, & 0.5 \\ 0.3, & 0.9, & 0.4 \\ 0.1, & 0.6, & 0.8 \end{bmatrix}$$

Quantum Max Pooling (2 × 2 window):

We slide a 2×2 window over the input feature map, and for each window, we select the maximum value and create the downsampled feature map.

For the first window, the values are [0.7, 0.2, 0.3, 0.9], and the maximum value is 0.9.

For the second window, the values are [0.2, 0.5, 0.9, 0.4], and the maximum value is 0.9.

For the third window, the values are [0.3, 0.9, 0.1, 0.6], and the maximum value is 0.9.

For the fourth window, the values are [0.9, 0.4, 0.6, 0.8], and the maximum value is 0.9.

The downsampled feature map after quantum max pooling is:

Pooled Feature Map (Quantum State):

$$\begin{bmatrix} 0.9, & 0.9 \\ 0.9, & 0.9 \end{bmatrix},$$

In this example, the spatial dimensions of the input feature map are reduced by half after applying quantum max pooling. The feature map retains the maximum values in each 2×2 window, capturing the most salient features of the input data.

Quantum pooling layers are crucial in QCNNs for downsampling feature maps obtained from quantum convolutional layers. By leveraging quantum operations, quantum pooling helps reduce the spatial dimensions of the data while retaining essential features, making the QCNN more computationally efficient. Quantum pooling is a critical step in quantum image recognition and pattern recognition tasks, allowing QCNNs to process and analyze image data in a quantum-enhanced manner.

10.3.2.3 Fully Connected Quantum Layers

Fully connected quantum layers are an integral part of QCNNs and represent the final stage of feature processing before making predictions in quantum-enhanced machine learning tasks. These layers are responsible for high-level reasoning and decision-making based on the extracted features from previous layers. In fully connected quantum layers, quantum operations are applied to process the feature representations, which are then used to make predictions for the given input data.

Let's elaborate on fully connected quantum layers with an example.

Example: Quantum Classification Task

Suppose we have a quantum image classification task where the goal is to classify images of handwritten digits into one of ten classes (0 to 9).

1. ***Quantum data encoding***: The input grayscale image of a handwritten digit is first encoded into quantum states using quantum data encoding techniques, such as amplitude encoding or quantum circuit encoding. For simplicity, let's consider amplitude encoding, where each pixel value in the image is mapped to the probability amplitude of a qubit.

   For example, let's consider a 4×4 grayscale image of the digit "3":

$$\text{Image} = \begin{bmatrix} 0, & 0, & 0, & 0 \\ 0, & 1, & 1, & 0 \\ 0, & 1, & 1, & 0 \\ 0, & 0, & 0, & 0 \end{bmatrix}.$$

   The quantum state representing this image can be written as:

$$|\psi\rangle = \alpha|0000\rangle + \beta|0100\rangle + \beta|0010\rangle + \gamma|0110\rangle.$$

   Here, α, β, and γ are complex probability amplitudes.

2. ***Fully connected quantum layers***: After quantum convolutional and quantum pooling layers have processed the image data, the feature map is flattened into a one-dimensional vector. This vector is then passed through the fully connected quantum layer to perform high-level reasoning.

Mathematical Example:
Let's assume that the flattened feature vector size is 16, and the fully connected quantum layer has three quantum gates (U1, U2, and U3). The quantum circuit can be represented as follows:

$$\text{Quantum Circuit} : \text{U3}\Big(\text{U2}\Big(\text{U1}\big(|\psi\rangle\big)\Big)\Big),$$

where U1, U2, and U3 are parameterized quantum gates that perform specific transformations on the input quantum state.

The quantum gates U1, U2, and U3 have adjustable parameters, which need to be trained using quantum circuit training techniques, such as variational algorithms.

3. ***Quantum circuit training***: Quantum circuit training involves optimizing the parameters of the quantum gates U1, U2, and U3 to minimize a cost function, which measures the discrepancy between the model's predictions and the true labels. Various optimization methods, such as gradient-based methods or stochastic optimization, are used to update the parameters during training.
4. ***Prediction***: Once the quantum circuit is trained, the final quantum state output by the fully connected quantum layer is measured to obtain the classical probabilistic outputs. These outputs represent the probabilities of the input image belonging to each class (0 to 9). The class with the highest probability is then selected as the predicted class for the input image.

Fully connected quantum layers play a crucial role in QCNNs. They process the feature representations obtained from quantum convolutional and pooling layers, enabling high-level reasoning and decision-making for quantum-enhanced machine learning tasks. By leveraging quantum parallelism and quantum states, fully connected quantum layers contribute to the potential efficiency and scalability of quantum machine learning models. Proper training of the quantum circuits in these layers is essential to achieve accurate predictions and harness the power of quantum computing for various classification and pattern recognition tasks.

10.3.2.4 Quantum Circuit Training

Quantum circuit training is a crucial step in training QNNs, including QCNNs. It involves optimizing the parameters of quantum circuits to learn relevant patterns and features from the input data. Quantum circuit training is analogous to classical deep learning's backpropagation, where gradients are used to update the model's parameters. Let's delve into some concepts and examples of quantum circuit training.

1. **Concepts of Quantum Circuit Training:**
   - ***Quantum neural networks***: QNNs are a class of quantum machine learning models that leverage quantum circuits and quantum gates to perform computations. These quantum circuits, similar to classical neural networks, contain parameters that need to be optimized during training.

- ***Cost function***: In quantum circuit training, a cost function is defined to quantify the difference between the predicted output of the quantum neural network and the actual output (ground truth). The goal is to minimize this cost function to train the quantum neural network effectively.
- ***Variational quantum algorithms***: Variational quantum algorithms (VQAs) are a class of quantum algorithms used for quantum circuit training that use variational techniques to optimize the parameters of the quantum circuits iteratively.

2. **Quantum Circuit Training Process:**
   - ***Quantum data encoding***: The input classical data is encoded into quantum states using quantum data encoding techniques. For example, amplitude encoding, or quantum circuit encoding can be used to map classical data into quantum states.
   - ***Quantum circuit initialization***: The quantum circuits are initialized with random values for their parameters. These parameters are the variables that will be optimized during training to learn meaningful representations from the input data.
   - ***Forward propagation***: The input quantum states are passed through the quantum circuit, and the quantum gates perform quantum operations on the quantum states. This process generates the output quantum state, which represents the prediction of the quantum neural network for the given input data.
   - ***Cost function evaluation***: The output quantum state is measured, and the result is compared with the actual output (ground truth) to compute the cost function. The cost function quantifies the discrepancy between the predicted output and the desired output.
   - ***Backpropagation and parameter updates***: The goal of quantum circuit training is to minimize the cost function. To achieve this, gradients with respect to the parameters of the quantum circuits are computed using techniques like parameter-shift rule or quantum automatic differentiation. These gradients indicate how the cost function changes with respect to changes in the parameters. Then, the parameters are updated using classical optimization techniques, such as gradient descent or Adam, to move towards the optimal parameter values that minimize the cost function.
   - ***Iterative training***: Quantum circuit training is an iterative process. The forward propagation, cost function evaluation, backpropagation, and parameter updates are performed repeatedly on batches of input data until the cost function converges to a minimum, indicating that the quantum neural network has learned the desired representations.
3. **Example of Quantum Circuit Training:** Let's consider an example where we want to train a simple quantum neural network to perform quantum image recognition. The input data consists of quantum states that represent grayscale images.
   - ***Quantum data encoding***: The grayscale pixel values of the input image are mapped to the probability amplitudes of qubits using amplitude encoding. For example, a pixel value of 0 is mapped to $\sqrt{0.1}|0\rangle + \sqrt{0.9}|1\rangle$.

- ***Quantum circuit initialization***: The quantum circuits are initialized with random values for the parameters (angle values) of quantum gates.
- ***Forward propagation***: The input quantum state is passed through the quantum circuit, and quantum gates perform operations on the quantum state to extract features from the image.
- ***Cost function evaluation***: The output quantum state is measured, and the result is compared with the ground truth (actual class label of the image) to compute the cost function.
- ***Backpropagation and Parameter Updates***: Gradients with respect to the parameters of the quantum circuit are computed using quantum automatic differentiation. The parameters are then updated using classical optimization techniques like gradient descent to minimize the cost function.
- ***Iterative training***: The forward propagation, cost function evaluation, backpropagation, and parameter updates are performed iteratively on batches of input quantum states until the cost function converges to a minimum, indicating that the quantum neural network has learned to recognize quantum images effectively.

Quantum circuit training is a fundamental process in training QNNs, enabling them to learn relevant features and patterns from input quantum states. By optimizing the parameters of quantum circuits using variational quantum algorithms, quantum circuit training allows quantum machine learning models to perform complex computations and solve various tasks efficiently. As quantum computing technology advances, quantum circuit training is expected to play a crucial role in unlocking the potential of quantum-enhanced machine learning and artificial intelligence.

10.3.3 Applications of Quantum Convolutional Neural Networks

Potential applications of QCNNs span various domains, benefiting from the quantum parallelism and quantum data encoding capabilities that QCNNs offer. These quantum-enhanced models have the potential to outperform classical CNNs in specific tasks due to their ability to process and represent data using quantum states.

1. ***Quantum image recognition***: QCNNs can be applied to image recognition tasks, where quantum states are used to encode image data, and quantum operations are performed to identify objects or patterns in images. Quantum image recognition may offer advantages in feature extraction and pattern detection, leading to improved accuracy and reduced computational complexity compared to classical approaches.
   Example: Classifying medical images, such as MRI scans, to detect specific diseases or abnormalities.
2. ***Quantum pattern recognition***: In addition to image recognition, QCNNs can be used for pattern recognition tasks, such as analyzing quantum states or quantum data structures. Quantum pattern recognition can have applications in quantum information processing and quantum cryptography.

Example: Identifying specific quantum states in quantum computing algorithms or recognizing quantum entanglement patterns.

3. ***Quantum image generation***: QCNNs have the potential to be used for quantum image generation tasks, where quantum states are manipulated to create new quantum image representations. Quantum image generation can be applied to quantum data compression and quantum art.
   Example: Generating quantum representations of classical images using quantum superposition and entanglement.
4. ***Quantum image super-resolution***: This is the process of enhancing the resolution of an image. Potentially, QCNNs can enhance image resolution by leveraging quantum properties to process and manipulate image data.
   Example: Upscaling low-resolution quantum images to higher resolution for improved analysis or visualization.
5. ***Quantum image denoising***: Image denoising aims to remove noise from images and improve their quality. Therefore, QCNNs can be utilized for quantum image denoising, where quantum operations are employed to filter out noise and enhance image clarity.
   Example: Denoising quantum images captured in noisy quantum computing experiments.
6. ***Quantum feature extraction***: Feature extraction is a critical step in many machine learning tasks and QCNNs can potentially offer advantages in extracting meaningful features from quantum data and quantum images.
   Example: Extracting relevant features from quantum states in quantum chemistry applications or quantum data analysis.
7. ***Quantum image segmentation***: Image segmentation involves partitioning an image into multiple regions or objects, so QCNNs can be applied to quantum image segmentation tasks to efficiently analyze and classify image regions.
   Example: Segmenting quantum images to identify distinct quantum phenomena or quantum objects.
8. ***Quantum content-based image retrieval***: Content-based image retrieval aims to find similar images based on their visual content. Thus, QCNNs can potentially enhance the efficiency and accuracy of content-based image retrieval systems in quantum domains.
   Example: Searching for similar quantum images based on their quantum representations.

The potential applications of QCNNs are diverse, ranging from image recognition and generation to pattern recognition and feature extraction in quantum domains. As quantum computing technology continues to advance, QCNNs are expected to play a significant role in quantum-enhanced machine learning and artificial intelligence. These applications represent exciting frontiers in quantum research, offering novel solutions to complex problems across various fields, including quantum information processing, quantum chemistry, and quantum image analysis.

10.4 SUMMARY

Deep quantum neural networks (QDNNs) represent a powerful fusion of quantum computing and deep learning techniques, combining the principles of quantum superposition and entanglement with the depth and expressiveness of classical deep neural networks. They consist of quantum neurons, quantum gates, and quantum circuits that process classical and quantum data for various applications. Quantum convolutional neural networks (QCNNs) are a specific type of QDNNs designed for computer vision tasks, utilizing quantum convolutional layers, quantum pooling layers, and fully connected quantum layers to process and analyze quantum image data. The architecture of QCNNs enables efficient feature extraction, pattern recognition, and image generation in quantum domains, offering potential advantages over classical approaches. Quantum circuit training is crucial for training QDNNs, employing variational quantum algorithms to optimize the parameters of quantum circuits. The potential applications of QCNNs include quantum image recognition, pattern recognition, image generation, and quantum image denoising, demonstrating their promise in quantum information processing, quantum chemistry, and beyond. As quantum computing technology advances, the exploration of deep quantum neural networks continues to be an exciting frontier in quantum machine learning and artificial intelligence.

BIBLIOGRAPHY

Biamonte, J., Wittek, P., Pancotti, N., Rebentrost, P., Wiebe, N., & Lloyd, S. (30 Nov 2017). Quantum Neuron: an elementary building block for machine learning on quantum computers. arXiv preprint arXiv:1711.11240.

Cong, I., Choi, S., Lukin, M. D., & Duan, L. M. (2019). Quantum convolutional neural networks. *Nature Communications*, 10(1), 1–9.

Farhi, E., Goldstone, J., & Gutmann, S. (2014). A quantum approximate optimization algorithm. arXiv preprint arXiv:1411.4028.

Fösel, T., Huembeli, P., Berenz, V., Jurcevic, P., Mezzacapo, A., & Woerner, S. (14 Apr 2020). Quantum Circuit Learning. arXiv preprint arXiv:2004.06755.

Harris, R., Johnson, M. W., Bunyk, P., Tolkacheva, E., Altomare, F., Berkley, A. J., ... Oh, T. (2018). Phase transitions in a programmable quantum spin glass simulator. *Science*, 361(6398), 162–165.

Hubregtsen, T., et al. (2019). Quantum convolutional neural networks. arXiv preprint arXiv: 1904.04767.

King, J., Yarkoni, S., Nevisi, M. M., Lam, R., & Hilton, J. P. (2015). Benchmarking a quantum annealing processor with the time-to-target metric. *Physical Review A*, 91(4), 042314.

Mari, A., et al. (2020) Transfer learning in hybrid quantum-classical neural networks. *Quantum Science and Technology*, 5(3), 250–262.

Peruzzo, A., et al. (2014). A variational eigenvalue solver on a photonic quantum processor. *Nature Communications*, 5, Article number: 4213.

Rønnow, T. F., Boixo, S., Isakov, S. V., Wang, Z., Wecker, D., Lidar, D. A., & Martinis, J. M. (2014). Quantum annealing with more than one hundred qubits. *Physical Review Letters*, 112(5), 057902.

Schuld, M., Bocharov, A., & Svore, K. M. (Wed, 24 Sep 2014). Supervised learning with quantum enhanced feature spaces. arXiv preprint arXiv:1409.6958.

Schuld, M., Sinayskiy, I., & Petruccione, F. (2015). The quest for a quantum neural network. *Quantum Information Processing*, 15(11), 1–22.

11 Ensembles and QBoost

Hariharan Bagavathi Thevar,
Ratna Kumari Neerukonda,
Anupama Cholanayakanahalli Govinda Reddy,
and Siva Rathinavelayutham
SRM Institute of Science and Technology, India

11.1 ENSEMBLES

In traditional classical ensemble machine learning algorithms, the learning process in an ensemble takes numerous models into consideration. Following this, the predictions from each of those particular multiple models are then pooled to improve overall performance. Numerous learning techniques are used for classifications, such as voting and aggregation for regression-type issues. Ensemble learning often requires the employment of many models, which might enhance the overall system's complexity. This intricacy can make interpreting and understanding the ensemble's behavior more difficult, especially when working with large ensembles. Ensembles sometimes need more computational resources than training and deploying a single model. Training many models and aggregating their predictions can take time and resources, especially for big datasets [1].

Quantum computing principles such as superposition and entanglement can be used to improve data representation, processing, and analysis in quantum machine learning algorithms. Each member of the ensemble represents a conceivable quantum system state. If we have a collection of qubits, for example, each qubit can be in a superposition of the base states |0 and |1. The ensemble would be made up of various combinations of these qubit states, such as |00, |01, |10, and |11, if we take a 2-bit quantum computer [2]. Quantum base models or classifiers are built utilizing quantum circuits or quantum algorithms in quantum computing. These foundational models may be generated utilizing a variety of quantum approaches, including quantum support vector machines, quantum neural networks, and quantum variational algorithms. Each quantum base model is intended to handle quantum input and forecast quantum states. Quantum gates, superposition, and entanglement may be used in these models to encode and manipulate data, capturing quantum correlations and patterns. Individual base models are generally trained on various subsets of the training data or with different beginning parameters in ensemble machine learning methods.

Training quantum base models in a quantum computing environment entails optimizing quantum circuits or algorithms to minimize a given loss function. Quantum optimization techniques, such as quantum gradient descent or quantum variational algorithms, can be used to repeatedly update the parameters of the quantum base

DOI: 10.1201/9781003429654-15

models in order to find the best configuration that minimizes the loss function. The training procedure seeks quantum states that encode the patterns and correlations in the training data, allowing for accurate predictions. After training the quantum basis models, their predictions are pooled to generate an ensemble prediction. To aggregate the predictions from the quantum basis models, ensemble combination procedures can be used. Voting (e.g., majority voting or weighted voting), averaging (e.g., taking the average or weighted average of forecasts), and stacking (training a meta-model on the predictions of the base models) are common combination approaches. These combination tactics, which take into account the probabilistic character of quantum states and their measurements, may be tailored to handle quantum predictions. When new quantum data is supplied to the ensemble model, each quantum basis model separately analyzes the data and makes predictions. To obtain the final ensemble forecast, these individual predictions are merged using the chosen combination approach. The ensemble of quantum base models intends to capitalize on the variety and complementary characteristics of multiple quantum models, improving prediction accuracy, resilience, and generalization [3]. A qubit can be in a state of superposition, which means it can be in more than one state at the same time. A linear combination of the basic states can be used to express the superposition numerically. The linear combination's coefficients would indicate the amplitude probability of the system state assuming a certain state. Let the state in equation (11.1) represent an arbitrary state of a single qubit.

The coefficients α and β are complex integers that indicate the probability of amplitude of $|\psi\rangle$.

$$|\psi\rangle = \alpha|0\rangle + \beta|1\rangle \tag{11.1}$$

Thus, an ensemble of classifiers is a classification approach that combines the output of many classifiers to get a final response. For instance, consider a binary classification issue with the set of classes $Y = \{-1, 1\}$. Let $E = a_1, a_2, a_3, \ldots a_{n-1}$ denote the collection of classifiers. Given an unknown data sample x, the output of each classifier is merged in a total to get the ensemble's final response, as stated in equation (11.2) [4].

$$\hat{y} = \text{sign}\left(\sum_{a \in E} w_\alpha f(\tilde{x}, a)\right) \tag{11.2}$$

where the sign is the sign function and $\hat{y}$ is the ensemble's answer (class).

ALGORITHM 1 PSEUDOCODE FOR ENSEMBLE QUANTUM MACHINE LEARNING

Quantum training data (quantum states and labels) are used as input.

1. Begin by creating an empty ensemble of quantum basis models.
2. Divide the quantum training data into m subsets.

3. Perform the following for $i = 1$ to m:
4. Develop a quantum base model (quantum circuit or algorithm) Q_i.
5. Using quantum optimization techniques, train the quantum base model Q_i on a subset of quantum training data.
6. Include the ensemble's learned quantum base model Q_i.
7. Create an empty array P of size n, where n is the number of quantum test data points.
8. Perform the following for $j = 1$ to n do:
9. Create an empty array P_j of size m.
10. if $i = 1$ to m do
11. Use the quantum base model Q_i to construct a prediction p_{ij} from the quantum test data point x_j
12. Put the prediction p_{ij} into the array P_j.
13. To obtain the final ensemble prediction p_j, apply the ensemble combination technique (e.g., majority voting, averaging, or stacking) to the predictions P_j.
14. Save the final ensemble prediction p_j to the array P.

Output: The final ensemble predictions for the quantum test data points are stored in array P.

In the pseudocode above, Q_i represents the quantum base model associated with the i-th subset of the quantum training data. P_j is an array of predictions for the j-th quantum test data point produced by each quantum base model Q_i. To produce the final ensemble prediction p_j, the ensemble combination approach is used to the predictions P_j.

11.2 QBOOST

Boosting is a machine learning ensemble approach used in traditional computing that combines a number of weak learners (usually decision trees) to produce a powerful prediction model. The primary principle underlying boosting is to train models progressively, with each new model concentrating on cases where the older models misclassified or made many inaccurate predictions. The first weak learner in the boosting process is taught using the complete training dataset. In each case, it produces predictions, however, initially these predictions might not be correct. In the training dataset, a starting weight is given to each sample. Most of the time, all weights are initially set evenly. The weights, however, are modified as the boosting process advances based on the accuracy of the earlier weak learners.

Boosting algorithms have the predisposition to overfit the training data, particularly if the weak learners are very complicated or the number of iterations is excessive. Overfitting happens when a model gets overly specialized in the training data and fails to generalize successfully to new data. Boosting is susceptible to noise in the training set or outliers. Because boosting methods give more weight to misclassified cases, noisy data can have a considerable influence on model performance, resulting in unsatisfactory outcomes [5].

The AdaBoost is a boosting algorithm where several complications affect the AdaBoost algorithm's computing needs and performance. AdaBoost necessitates several cycles of training weak learners. The temporal complexity of each iteration is determined by the weak learner algorithm used. In this part, we apply quantum approaches to increase AdaBoost's complexity. We divide our quantum boosting algorithm into phases. The majority of the technical work in our quantum boosting approach is focused on decreasing training errors [6].

1. In order to generate a weak hypothesis h_t under an approximation distribution D_t invoke the weak quantum learner A over the training set S.
2. Our technique computes ϵ_t^l, an approximation to $\tilde{\epsilon}_t = P_{\gamma_{x \sim \tilde{D}^t}[h_t(x) \neq C(x)]}$ by executing quantum queries to h_t.
3. Using ε_t calculate a weight α_t. Output an after-T steps $H(x)$ hypothesis = sign $\sum_{t:1}^{T} \alpha_t h_t(x)$.

Quantum boosting algorithms use the capabilities of quantum computers to enhance the precision and effectiveness of traditional boosting methods. Weak learners in classical boosting are frequently decision trees or other traditional machine learning models. The weak learners in quantum boosting are quantum circuits or quantum models that can run on quantum computers. Quantum weak learners, which are quantum circuits or models that outperform conventional machine learning algorithms, are the foundation of quantum boosting techniques. Quantum gates and quantum processes are employed in the implementation of these weak learners. A quantum dataset is used to train the first quantum weak learner. The quantum dataset is created either by directly utilizing quantum data or by encoding classical data into quantum states. On the basis of the quantum dataset, the first quantum model produces predictions, however, these predictions may initially be incorrect [7].

The quantum dataset gives each case a starting weight. Usually, their weights are originally set evenly. But when quantum boosting advances, the weights are changed in accordance with the precision of the earlier quantum weak learners. Examples that are incorrectly categorized are given more weight, highlighting their significance for further revisions. Boosting comprises a number of iterations, each of which aims to enhance the predictions made by the preceding quantum weak learners. A fresh quantum weak learner is taught on the quantum dataset in each cycle. However, the weights of the instances are changed throughout training to highlight the incorrectly categorized cases.

When fresh quantum weak learners are taught, their predictions are integrated with the predictions of prior learners using a weighted voting method. The weight of each quantum weak learner is determined by its accuracy in categorizing the instances. Various approaches, such as quantum state superposition or quantum amplitude amplification, can be used to combine quantum predictions.

After all of the quantum weak learners have been taught and merged, the final prediction is formed by aggregating all of the quantum weak learners' predictions. To compute the final prediction, the weights of individual quantum weak learners are taken into account.

ALGORITHM 2 PSEUDOCODE FOR QBOOST ALGORITHM

Begin:

- A collection of quantum weak learners is designated by the letters Q_1, Q_2, Q_3 and Q_T, where T is the total number of weak learners.
- Assign starting weights to the quantum dataset's training instances. $D = (x_1, y_1), (x_2, y_2)$ and (x_n, y_n) denotes the quantum dataset, where x_i is the input quantum state and y_i is the associated binary label.

Iteration:
For each t = 1 to T iteration:

- Train the t-th quantum weak learner Q_t on the weighted quantum dataset D_t, where D_t is a modified version of D based on example weights.
- Determine the _t error of Q_t on the quantum dataset D_t.
- Based on the mistake _t, compute the weight _t associated with Q_t. Q_t's contribution to the final forecast is quantified by the weight _t.
- For the following iteration, adjust the weights of the samples in the quantum dataset D_t to emphasize the misclassified examples.

End:

- Using a weighted voting mechanism, combine the predictions of all quantum weak learners. $F(x) = \text{sign}\,(\Sigma_t \alpha_t Q_t(x))$ is the ultimate prediction for a given input quantum state x.
- The sign function turns the total to a binary prediction in this formula, and $Q_t(x)$ is the prediction of the t-th quantum weak learner for the input quantum state x.

11.3 QUANTUM ANNEALING

Quantum annealing is a quantum computing approach for determining the best solution to a problem. It takes advantage of quantum mechanical phenomena such as superposition, entanglement, and quantum tunneling. It is analogous to annealing in material science. Instead of boosting temperatures, the energy of qubits is increased, and a lower energy state is gradually obtained. It presents a method for solving NP-Hard issues. The quantum annealer is a probabilistic rather than a deterministic algorithm. The computer provides several replies in a short period of time (in microseconds). We not only receive the greatest answer but also several additional excellent options from which to choose.

Quantum annealing is a method for solving optimization issues that makes use of quantum mechanical concepts. It is concerned with determining the lowest energy state (minimum) of a particular objective function or cost function. The quantum computers developed by D-Wave are intended to tackle optimization issues using quantum annealing techniques. Their quantum processing units (QPUs) are made up of superconducting qubits that are designed to execute quantum annealing procedures. The quantum computers developed by D-Wave are built on superconducting qubits that can represent quantum states. The qubits are modified to encode the problem and seek for the best solution. The quantum annealers developed by D-Wave are specially designed to handle combinatorial optimization challenges. The quantum annealers developed by D-Wave are specially designed to handle combinatorial optimization challenges.

In quantum computing, quantum annealing uses mathematical language to express the ideas and processes involved [8]. The following are some significant mathematical notations used in quantum annealing.

Qubits are the basic building blocks of quantum information. Dirac's bra-ket notation is used to express them, with a qubit state marked as $|\psi\rangle$. A two-qubit system, for example, can be written as $|\psi\rangle = \alpha|00 + \beta|01 + \gamma|10 + \delta|11$, where α, β, γ and δ are complex coefficients. The Hamiltonian is a mathematical operator that expresses a quantum system's entire energy. It is commonly abbreviated as H and is made up of two terms: the problem Hamiltonian (H_p) and the driver Hamiltonian (H_d).

$H = AH_p + BH_d$ is the Hamiltonian, where A and B are coefficients that regulate the respective intensities of the two terms. The time-dependent development of the Hamiltonian during the annealing process is determined by the annealing schedule. It is frequently expressed as $s(t)$, where t is the time parameter. The annealing schedule defines how the Hamiltonian coefficients A and B change over time. The purpose of quantum annealing is to discover the ground state or configuration with the lowest energy. The energy of a quantum state is symbolized by the symbol $E(\psi)$ or $\langle\psi|H|\psi\rangle$, where H represents the Hamiltonian. The goal is to discover the state $|\psi\rangle$ that uses the least amount of energy [9].

Consider the "Traveling Salesman Problem" (TSP) as a basic numerical illustration of how quantum annealing may be used to tackle optimization issues in quantum computing. The TSP entails determining the shortest path that visits a collection of cities and returns to the beginning city without visiting any cities twice. Assume we have four cities labeled A, B, and C, and we want to discover the shortest path that visits all of them and returns to the starting point. A distance matrix can be used to express the problem:

	A	B	C
A	0	5	8
B	5	0	6
C	8	6	0

To solve this issue using quantum annealing in a quantum computing context, we must first transfer it to a quantum Hamiltonian and then run the annealing process. However, it is crucial to remember that existing quantum technology may not have enough qubit counts or coherence times to solve real-world TSP cases. Here's a simplified example of how to utilize quantum annealing to solve a problem for a smaller TSP instance with three cities (A, B, and C). In the TSP, define the goal function, which is the total distance traveled. Let's write D(A, B), D(B, C), and D(C, A) to indicate the distance between cities [9]. To encode the goal function, build the

Hamiltonian. The Hamiltonian may be written as $H = D(A, B)\ |AB| + D(B, C)\ |BC| + D(C, A)\ |CA|$, where |A, |B, and |C represent the quantum states associated with cities A, B, and C, respectively. Use the quantum annealing process to progressively convert the Hamiltonian from an initial to a final state. The annealing schedule governs the system's development, allowing it to locate the ground state corresponding to the shortest path. Measure the final state of the quantum system and extract the solution after the annealing process. The measurement data will show the cities in the shortest route order. It is crucial to note that the offered example is simplified, and the mapping of the TSP to a quantum Hamiltonian might be more difficult for bigger issue cases [10].

11.4 QUADRATIC UNCONSTRAINED BINARY OPTIMIZATION (QUBO)

The important terms related to quantum computers terms are binary quadratic models because these are the only language quantum computers understand. When you present an issue or submit it to a quantum processing unit, you must first transform it into a binary quadratic model, which can then be processed by the quantum processing unit to give a decent result. The binary quadratic model is made up of three words. The first one is binary which stands for two states or two variables [11].

$$q(x,y) = ax^2 + bxy + cy^2$$

As you can see, we have a function called Q that is a function of x and y, so there are x and y values that are then dependent on Q. Thus, as the value of x and y change, the value of Q changes, so Q is called a dependent variable, and x and y are independent variables.

Quadratic unconstrained binary optimization (QUBO) is a mathematical framework for modeling and solving combinatorial optimization issues. The issue is represented as a quadratic objective function with binary variables. The purpose of QUBO is to identify the binary variable assignments that minimize (or maximize) the quadratic objective function while meeting all constraints. The binary variables can have values of 0 or 1.

Consider the following basic optimization problem which is a constrained optimization:

Reduce the quadratic function $f(x) = x^2 - 4x + 4$ its simplest form.
Subject to the restriction x as 0, 1

We may frame this problem as a quadratic binary optimization problem by introducing a binary variable y that represents the variable x in order to solve it using quadratic binary optimization. The following transformation can be used:

$$x = y(1-y)$$

Now, let's convert the objective function and constraint into their binary forms:
Objective function:

$$f(x) = x^2 - 4x + 4$$

$$= (y(1-y))^\wedge 2 - 4(y(1-y)) + 4$$

$$= y^\wedge 2 - 2y^\wedge 2 + y^\wedge 2 - 4y + 4$$

$$= -2y^\wedge 2 - 4y + 4$$

Constraint:

$$x \in \{0,1\}$$

$$=> y(1-y) \in \{0,1\}$$

The original problem has now been turned into a quadratic binary optimization problem. It can be expressed in the following standard form:

$$\text{Reduce: } -2y^\wedge 2 - 4y + 4$$

$$\text{Subject to } y(1-y) € \{(0,1)\}$$

Solving this quadratic binary optimization issue will provide us with the optimal y value, which we can then use to get the equivalent x value. Here are several QUBO problems that can be handled with quantum computing. One of the major problems is the Max-cut problem which is solved by a QUBO problem in the following way.

11.4.1 The Max-Cut Problem

Given an undirected graph containing a set of nodes and edges, the goal is to divide the nodes into two disjoint sets while maximizing the number of edges between the two sets. This problem may be expressed as a QUBO problem, with the variables representing node assignment to sets and the objective function aiming to maximize the number of severed edges.

The following is a definition of the Max-Cut problem:

Given an undirected graph G = (V, E), we wish to discover a partition of V into two disjoint sets, S and S', such that the number of edges between S and S' (i.e., the cut) is maximized [12].

We can use quantum computing to solve the Max-Cut issue by mapping it to a QUBO problem.

The QUBO formula is as in equation (11.3):

$$\text{Minimize: } H = -\Sigma(i,j) \in E\ w(i,j)\left(x_i * (1-x_j) + (1-x_i) * x_j\right) \quad (11.3)$$

where:

(i, j) represents a graph edge.

The weight of the edge between nodes i and j is represented by G. $w(i, j)$.

x_i and x_j are binary variables that indicate whether node i and j are assigned to set S or S′.

The goal is to minimize the above Hamiltonian H, which equates to having the most clipped edges. Quantum computing technologies such as quantum annealing, or variational quantum algorithms can be used to solve this QUBO formulation. Quantum annealing searches for the lowest energy state of the QUBO problem, which corresponds to the optimum node partition. In contrast, variational quantum algorithms employ parameterized quantum circuits to iteratively optimize the QUBO objective function, eventually convergent to the ideal solution.

We may explore the solution space more efficiently and discover better answers than classical optimization techniques in some situations by transferring the Max-Cut problem to a QUBO formulation and employing quantum computing technologies.

11.5 ISING MODEL

The Ising model is a traditional statistical physics model for studying magnetic systems. It represents a collection of spins that can be in either an "up" or "down" state and their interactions are guided by an energy function that relies on their respective orientations. The Ising model has also been utilized as a starting point for understanding some types of quantum systems in the context of quantum computing. The Ising model is especially relevant to certain types of quantum annealing devices, such as those based on superconducting qubits. It has a lattice structure because it only considers the closest neighbor interactions dictated by the alignment or anti-alignment of spin projections along the Z axis. The alignment in the Ising model is likewise regulated by the external magnetic field [13].

The Hamiltonian of the quantum Ising model is given by equation (11.4):

$$H = -J\,(\overset{\Sigma}{(i,j)} z_i z_j + g \sum_j x_j \tag{11.4}$$

The indexes i, and j are representing lattice sites and the sum is taken over the pairs of nearest neighbors. The prefactor j has dimensions of energy and factor g indicates the strength of the externally applied field relating to nearest neighbor interaction. The x_j and z_j represent components or spin algebra acting over spin variables of corresponding sites.

The 1D model of each lattice site is a two-dimensional complex Hilbert space. The Hamiltonian of the 1D quantum model poses z_2 symmetry, it remains invariant under the transformation of flipping of all the spins about the Z axis. As we know that every phase transition involves some kind of spontaneous symmetry breaking. The 1D model also has two phases depending on whether the ground state preserves or breaks the spin-flip symmetry [14]. Qubits in the Ising model are often expressed using Dirac notation or ket notation. A qubit can be expressed as a column vector:

$$|\psi\rangle = \alpha|0\rangle + \beta|1\rangle,$$

where α and β are complex probability amplitudes, $|0\rangle$ represents the "0" state, and $|1\rangle$ represents the "1" state.

11.6 SPARSITY, BIT DEPTH, AND GENERALIZATION PERFORMANCE

Sparsity, bit depth, and generalization performance are key concerns in the context of quantum computing, as they relate to the efficiency and efficacy of quantum algorithms. Sparsity is a quality of a quantum system, or the data on which it acts, in which a large part of the components or parameters are zero or near to zero. Sparsity denotes that only a tiny proportion of the components in a quantum state or a quantum operation are non-zero. Quantum systems are distinguished by short coherence durations, imprecise gates, and vulnerability to noise and mistakes. Using sparsity to limit the number of processes and resources required to conduct a computation can result in increased efficiency. Quantum algorithms can minimize the number of operations, gate operations, or measurements required by focusing on non-zero items and disregarding zero elements, reducing the effect of mistakes and the total computational cost. Sparsity is also important in quantum computing data representation and transmission. In terms of memory consumption and communication bandwidth, storing and sending sparse quantum states or processes can be more efficient. Rather than recording and transmitting all of the constituents of a quantum state, just the non-zero elements can be represented and sent, resulting in decreased storage needs and enhanced communication efficiency [15]. Sparsity in quantum computing can be theoretically described using various notations and frameworks depending on the circumstance. The following are some typical mathematical notations for describing sparsity in quantum computing:

$$|\psi\rangle = \Sigma c_i|i\rangle \tag{11.5}$$

where $|i\rangle$ represents a basis state, and c_i represents the coefficient or probability amplitude associated with that state. In the case of sparsity, only a few c_i values are non-zero, while the rest are zero or negligible. Sparsity can also be seen in quantum processes or transformations. Matrix or operator notation is used to express sparse quantum operators. A sparse quantum operator is often expressed in this scenario as a matrix with many zero members. For instance, if operator A is sparse, its matrix representation A_{ij} has a large number of zero entries. The number of quantum bits, or qubits, used to represent and process information in quantum computing is referred to as bit depth. Each qubit can exist in a state of superposition, representing both "0" and "1" at the same time. As a result, the bit depth of a quantum computing system dictates the amount of information that can be processed. The number of bits used to represent the color or intensity of a pixel in an image or the number of bits required to represent a sample in digital audio, is referred to as bit depth in classical computing.

A quantum bit, or qubit, is the fundamental unit of information in quantum computing. Unlike conventional bits, which can either be a 0 or a 1, qubits may exist in a superposition of both 0 and 1 at the same time due to quantum physics principles. One of the key qualities of qubits that permits quantum computers to do certain types of calculations more efficiently than conventional computers is their capacity to be in a superposition. A quantum computer's qubit count is an essential indicator of its computing capacity. In general, increasing the number of qubits increases the complexity and scale of issues that may be solved. The number of qubits, however, does not precisely correlate to the "bit depth" in the traditional sense. Instead, in quantum computing, the "bit depth" may be conceived of as the number of different states that the qubits can represent. The quantum computer's state space expands exponentially with each new qubit. For example, two qubits can represent four different states: 00, 01, 10, and 11. Eight states may be expressed with three qubits, and so on. One of the reasons quantum computers have the ability to tackle certain problems more effectively than classical computers is due to the exponential rise in the number of states [16].

In classical machine learning, generalization refers to a trained model's capacity to perform effectively on previously encountered data that was not utilized during the training process. It assesses the model's capacity to recognize and generalize patterns from training data in order to generate accurate predictions on new, previously unknown cases. Metrics including accuracy, precision, recall, or mean squared error are commonly used to assess generalization performance.

One strategy, similar to traditional machine learning, is to divide the given dataset into distinct training and test sets. The quantum model is trained on the training set before being tested on the test set for generalization performance. The test set provides data that the model did not view during the training phase, providing an assessment of the model's ability to generalize to previously unknown data. Cross-validation is another approach employed in conventional machine learning that can be applied to quantum machine learning models. Cross-validation divides the dataset into numerous subsets, or folds, and trains and evaluates the model multiple times, with each fold acting as the test set once, and the remaining folds utilized for training. By utilizing diverse subsets of the data, this technique allows for a more robust evaluation of generalization performance. Overfitting is a widespread problem in machine learning, especially quantum machine learning. It happens when a model grows too complicated and absorbs too much noise or too many particular characteristics from the training data, resulting in poor generalization to fresh data. Regularization strategies, such as including regularization terms into the quantum model or employing specialized optimization methods, can aid in mitigating overfitting and improving generalization performance [17].

11.7 MAPPING TO HARDWARE

Quantum computing is the application of quantum mechanical phenomena to computation. Unlike traditional computers, which use bits to encode and process data, quantum computers employ quantum bits, or qubits, which may exist in several states at the same time due to a feature known as superposition. These qubits are manipulated and controlled by the hardware components in quantum computing systems. The fundamental building element of quantum computers is qubits. Because of superposition, they can represent both 0 and 1, allowing quantum computers to do parallel operations. Superconducting circuits, trapped ions, topological qubits, and silicon-based qubits are examples of common qubit implementations [18].

Quantum gates, which are similar to conventional logic gates, are used to manipulate qubits. These gates can conduct superposition, entanglement, and quantum interference. The Hadamard gate, CNOT gate, Pauli gates (X, Y, Z), and Toffoli gate are all examples of quantum gates. A quantum register is a group of qubits used to store and modify quantum information. It is the quantum counterpart to a classical register. Complex quantum processes can be accomplished by entangling qubits within a register. Quantum computers frequently run at temperatures close to absolute zero (about −273°C or −459°F). Cryogenic equipment, such as dilution refrigerators or cryocoolers, are utilized to chill the qubits to superconducting temperatures, which reduces noise and increases coherence time.

Precision control over qubits and gates is required for quantum computers. Microwave and radiofrequency generators, control electronics, and timing circuits are all components of control systems. These systems generate the control signals required to operate and measure the qubits.

Physical qubit allocation, also known as qubit mapping, is the process by which quantum computers are transferred to hardware. The logical qubits of a quantum algorithm are assigned to the physical qubits of the available hardware in this procedure. The objective is to discover the best mapping that minimizes the number of physical qubits needed while still satisfying any connection limits imposed by the hardware design. A quantum circuit made up of logical qubits and quantum gates is used to describe the quantum algorithm. The logical qubits represent the algorithm's qubits, and the gates represent the operations that will be done on those qubits.

It is crucial to understand the individual hardware architecture's features and restrictions, such as the number and connection of physical qubits, available gate operations, error rates, and coherence times. This data is critical for translating logical qubits to physical qubits, and determining how logical qubits are assigned to physical qubits. This mapping should fulfil the hardware's connection limits while also taking into consideration elements like qubit connectivity, gate availability, and error rates.

After mapping the logical qubits to physical qubits, the quantum gates of the circuit are mapped to the hardware's accessible gate actions. This procedure entails finding the particular gate sequences or decompositions necessary to construct the desired gates using the hardware's available gates. The mapping may be improved further to enhance metrics like gate count, gate depth, qubit connection, and error prevention.

To minimize resource needs and increase the overall performance of the quantum circuit, techniques such as gate merging, gate swapping, and gate scheduling are used. The optimized quantum circuit produced by the mapping process is compiled into a set of instructions or control signals that may be performed on hardware. These instructions comprise gate pulses, qubit manipulation control signals, and measurement activities. After that, the compiled circuit is run on the physical quantum hardware. The process of mapping quantum algorithms to hardware is iterative and involves feedback loops. The performance of the mapped circuit is assessed on hardware, and the mapping may be modified or altered depending on the observed findings to solve any difficulties or restrictions identified during execution. This recurrent improvement contributes to the mapping process's overall efficiency and dependability [19].

11.8 COMPUTATIONAL COMPLEXITY

Quantum computational complexity theory extends conventional computational complexity theory by accounting for quantum computers' unique traits and capabilities, such as superposition, entanglement, and interference. The number of quantum gates required to conduct a computation is an essential metric for determining complexity. The depth of a quantum circuit, which indicates the number of gates applied in succession, is frequently employed as a temporal complexity metric. To perform efficient calculations, quantum algorithms try to minimize the depth of the circuit.

The space complexity of a quantum algorithm is determined by the number of qubits and the size of the quantum register required to solve a problem [20]. The scalability and resource needs of quantum algorithms are shown by the rise of qubits or quantum register size with input size.

One of the primary incentives for quantum computing is the possibility of quantum speedup, in which certain tasks may be solved much quicker on a quantum computer than on a conventional computer. To understand the possible speedup, complexity analysis in quantum computing sometimes entails comparing the resources required for classical algorithms with their quantum equivalents.

11.9 SUMMARY

Ensemble machine learning algorithms incorporate a learning process that examines and predicts several models to enhance overall performance. This intricacy can make understanding the ensemble's behavior challenging, especially when working with large ensembles.

Superposition and entanglement, two quantum computing concepts, can improve data representation, processing, and analysis in quantum machine learning algorithms. Each ensemble member represents a conceivable quantum system state, such as a collection of qubits in a base state superposition.

Quantum base models are meant to manage quantum input and anticipate quantum states utilizing quantum circuits or algorithms. These models may be optimized using quantum optimization techniques such as quantum gradient descent or quantum variational algorithms. Following training, predictions are pooled to provide a prediction. Voting, averaging, and stacking are examples of ensemble combination processes that aggregate predictions from quantum basis models. To handle quantum predictions, these strategies take into account the probabilistic nature of quantum states and their observations. The ensemble of quantum base models takes advantage of the diverse and complimentary properties of numerous quantum models, boosting prediction accuracy, robustness, and generalization.

Ensemble machine learning algorithms use a linear combination of basic states to express superposition numerically. The coefficients indicate the amplitude probability of the system state assuming a certain state. An ensemble of classifiers is a classification approach that combines the output of many classifiers to get a final response. For example, in a binary classification problem, the output of each classifier is merged to get the ensemble's final response.

Boosting is a machine learning ensemble approach that combines weak learners, such as decision trees, to produce powerful prediction models. It trains models progressively, focusing on cases where older models misclassified or made inaccurate predictions. The first weak learner uses the complete training dataset, producing predictions initially with uncertain accuracy. Boosting algorithms can overfit the training data, especially if the weak learners are complicated or the number of iterations is excessive. Noise in the training set or outliers can influence model performance, resulting in unsatisfactory outcomes.

The AdaBoost algorithm has several complications, requiring multiple cycles of training weak learners and increasing its complexity. Quantum approaches are applied to increase AdaBoost's complexity, dividing the algorithm into phases and focusing on decreasing training errors. Physical qubit allocation, also known as qubit mapping, is the process of transferring quantum computers to hardware. The objective is to find the best mapping that minimizes the number of physical qubits needed while satisfying connection limits imposed by the hardware design. A quantum circuit consists of logical qubits and quantum gates, with logical qubits representing the algorithm's qubits and gates representing the operations. The mapping process can be improved to enhance metrics like gate count, gate depth, qubit connection, and error prevention. Techniques like gate merging, gate swapping, and gate scheduling are used to minimize resource needs and increase overall performance. The optimized quantum circuit is compiled into instructions or control signals that can be run on physical quantum hardware.

Quantum computational complexity theory takes into account the unique properties and capabilities of quantum computers, such as superposition, entanglement, and interference. A critical parameter for calculating complexity is the number of quantum gates required for computing. A quantum circuit's depth is employed as a temporal complexity meter. Quantum algorithms strive to reduce circuit depth for more efficient computations. The number of qubits and the size of the quantum record required to solve a problem define space complexity. The major motivation for quantum computing is speed, with complexity analysis comparing conventional algorithms with quantum versions to understand possible speedup.

REFERENCES

[1] X Dong, Z Yu, W Cao, Y Shi, Q Ma. A survey on ensemble learning, *Frontiers of Computer Science*, 2020, 14: 241–258.

[2] J Singh, M Singh. Evolution in quantum computing, *International Conference Systems Modeling and Advancement in Research Trends (SMART)*, 2016.

[3] C Chen, D Dong, B Qi, Quantum ensemble classification: A sampling based learning control approach, *IEEE Transactions on Neural Networks and Learning systems*, 2016, 28, 1–32.

[4] I C S Araujo, A J da Silva, Quantum ensemble of trained classifier, *International Joint Conference on Neural Networks*, 2020, IEEE.

[5] R E Schapire, *The Boosting Approach to Machine Learning Algorithm: An Overview, Nonlinear Estimation and Classification*, Springer.

[6] S R Arunachalam, R Maity, *Quantum Boosting: Proceedings of the 37th International Conference on Machine Learning, PMLR*, 2020, 119, 377–387.

[7] X Wang, Y Ma, M Hsieh, M Yung, *Quantum Speedup in Adaptive Boosting of Binary Classification*, 2020, Science China Physics, Mechanics and Astronimy.

[8] C C McGeoch, R Harris, S P Reinhardt, *Practical annealing-based quantum computing, Computer*, 2019, 52(6), 38–46.

[9] C R Laumann, R Moessenenr, A Scardicchio, S L Sondhi, Quantum annealing: The fastest route to quantum computation? *The European Physical Journal Special Topics*, 2015, 224, 75–88.

[10] T Kieu, *The Traveling Salesman Problem and Adiabatic Quantum Computation: An Algorithm*, 2019, Quantum Information Processing.

[11] R H Warren, Solving the travelling salesman problem on quantum annealer, SN *Applied Sciences*, 2020, 2, 75.

[12] L Braine, D J Egger, J Glick, S Woerner, Quantum algorithms for mixed binary optimization applied to transaction settlement, *Quantum Engineering*, 2021, 2, 1–12.

[13] G G Guerreschi, A Y Matsuura, QAoA for max-cut requires hundreds of qubits for quantum speed-up, *Scientific Reports*, 2019, 9, 1–8.

[14] E C Bauckhage, K Brito, C O Cvejoski, R Sifa, S Wrobel, Ising models for binary clustering via adiabatic quantum computing, *Energy Minimization Methods in Computer Vision and Pattern Recognition*, 2018, Springer.

[15] G Pagano, A Bapat, P Becker, C Monroe, Quantum approximate optimization of the long-range Ising model with a trapped-ion quantum simulator, *National Academy of Sciences*, 2020, 117, 25396–25401.

[16] D Dhawan, M Metcalf, D Zqid, Dynamical self-energy mapping (DSEM) for creation of sparse Hamiltonians suitable for quantum computing, *Chemical Theory and Computation*, 2021, 17, 7622–7631.

[17] R Blume-Kohout, K C Young, A volumetric framework for quantum computer benchmarks, *Quantum Journal*, 2020, 4, 362.

[18] C Havenstein, D Thomas, S Chandrasekaran, Comparisons of performance between quantum and classical machine learning, *SMU Data Science*, 2018, 4. https://scholar.smu.edu/datasciencereview/vol1/iss4/11

[19] J Tejada, E M Chudnovsky, E del Barco, J M Hernandez, T P Spiller, Magnetic qubits as hardware for quantum computers, *Nano Technology*, 2001, 12, 1–12.

[20] J S Kottmann, M Krenn, T H Kyaw, S Alperin-Lea, A Aspuru-Guzik, Quantum computer-aided design of quantum optics hardware, *Quantum Sciences and Technology*, 2021, 6, 3–11.

12 Quantum Process Tomography and Regression

Kanaga Priya Palanisamy
Sri Eshwar College of Engineering, India

Thanga Revathi Shanmugakani
SRM Institute of Science and Technology, India

Gomathy Balasubramanian
PSG Institute of Technology and Applied Research, India

Reethika Anandan
Sri Ramakrishna Engineering College, India

12.1 CHANNEL-STATE DUALITY

Channels and states are crucial in defining the behavior and manipulation of quantum systems within the domain of quantum information theory. Channels represent the transformations that a quantum system can undergo, while states characterize the quantum information encoded within the system. The concept of channel-state duality arises from the deep interconnection between these two essential components of quantum mechanics.

12.1.1 Quantum Channels: Definition and Properties

A quantum channel is a mathematical representation of a physical process that acts on quantum systems. It describes how an input quantum state evolves into an output state after transforming. The mathematical representation of a quantum channel can be done by a completely positive trace-preserving (CPTP) mapping, which guarantees that probabilities are conserved and the changed state stays positive semidefinite. Quantum channels possess several notable properties. First, they can be both reversible and irreversible. Reversible channels are unitary transformations, meaning they can be undone by applying the inverse operation. On the other hand, irreversible channels are non-unitary and represent processes such as measurements or decoherence, where information is lost. Second, quantum channels can be categorized as

DOI: 10.1201/9781003429654-16

either noiseless or noisy, depending on their ability to preserve or degrade quantum information during the transformation.

12.1.2 Quantum States: Description and Manipulation

Quantum states describe the complete information about a quantum system. Vectors in a complex vector space, often the Hilbert space, are used to represent them. Quantum states exhibit unique properties such as superposition and entanglement, enabling quantum systems to perform computation and communication tasks beyond the capabilities of classical systems. Quantum states can be manipulated through various quantum operations, including unitary transformations, measurements, and state preparations. Unitary transformations are reversible operations that preserve the norm and inner product of the state vector. Measurements collapse of the state into one of the possible measurement outcomes, providing probabilistic information about the system. State preparations allow for the generation of specific quantum states by carefully engineering the system.

12.1.3 Channel-State Duality: Connecting Channels and States

The channel-state duality establishes a deep connection between quantum channels and states. This illustrates that the transformation of a quantum state can be equivalently described as the effect of a channel acting on the state. This duality implies that the description of a channel and the description of a state carry the same essential information about the quantum system. Mathematically, the duality between channels and states can be expressed through the concept of Choi-Jamiolkowski isomorphism. According to this isomorphism, every quantum channel corresponds to a unique quantum state, known as the Choi state, and vice versa. This correspondence enables us to study quantum processes by characterizing the associated states and vice versa, leading to powerful tools for understanding and manipulating quantum systems. The channel-state duality has practical implications in various areas of quantum information science. It serves as the cornerstone for quantum process tomography, which tries to rebuild quantum channels experimentally by characterizing the corresponding states. It also plays a crucial role in quantum error correction, where quantum channels are mitigated by encoding information in specially designed quantum states. Hence, in this section, the channel-state duality reveals the intimate relationship between quantum channels and states. Understanding this duality provides insights into the behavior and manipulation of quantum systems, leading to advancements in quantum information processing, quantum communication, and quantum computation. The following sections provide a comprehensive overview of quantum process tomography (QPT) which is used for quantum processing, group theory concepts for understanding about quantum physics, representation theory for understanding about algebraic structures, parallelism in quantum computing in QPT provides understanding about exponential speeding up over traditional computers, and finally, the criteria used to find the optimal state in QML tasks is discussed to provide a clear understanding of terminologies.

12.2 QUANTUM PROCESS TOMOGRAPHY (QPT)

Quantum process tomography (QPT) is a strong method for characterizing and comprehending quantum actions or processes. Reconstructing the complete description of a quantum channel allows us to gain insights into how quantum systems evolve, and to assess the quality of quantum operations. This section delves into the methodology and goals of QPT. It is a technique that can be used to characterize quantum gate implementation in experimental settings, especially when dealing with a small number of qubits. In addition, QPT is both a practical instrument for evaluating gate state transformation and a theoretical tool for understanding the impact of noise and imperfections on gate performance. While fidelity and distance metrics provide a single numerical indication of gate deviation from the ideal, QPT provides a more in-depth examination by offering detailed insights into the unique mistakes induced by distinct flaws. Qiskit and QuTiP are two separate software packages used in quantum computing and quantum physics, respectively. QuTiP (Quantum Toolbox in Python) is a Python-based quantum toolbox, and a free and open-source software program used to simulate the dynamics of open quantum systems. It is most commonly employed in quantum physics and quantum optics. QuTiP is developed in Python and offers a wide range of tools for solving quantum master equations and simulating quantum systems. QuTiP is intended for researchers and scientists interested in quantum mechanics. It can simulate quantum systems such as Hamiltonian dynamics, Lindblad master equations, and Monte Carlo wave function simulations. QuTiP is commonly used for activities such as quantum system modeling, quantum optics investigations, and quantum information theory research. QuTiP's capabilities in this field have recently been enhanced with the addition of support for the computation and visualization of quantum process tomography matrices. The quantum process tomography matrices for various qubit gates have been computed for C-NOT, SWAP, iSWAP, √iSWAP, $\pi/2$ phase gate, and S-NOT gates are shown in Figure 12.1. These are all illustrations of perfect quantum gates.

IBM's Qiskit is an open-source quantum computing framework. It contains a collection of tools and libraries for programming and dealing with quantum computers. Qiskit enables users to design, alter, and simulate quantum circuits before running them on actual quantum hardware offered by IBM's quantum cloud services. Qiskit is intended to aid with quantum computing research and development. It comprises components for designing quantum circuits, developing quantum algorithms, simulating quantum systems, and accessing quantum hardware. Qiskit is compatible with a wide range of quantum hardware, including superconducting qubit-based quantum computers. Terra is the core package of Qiskit, and it contains the essential building blocks required to program quantum computers. The quantum circuit is the fundamental unit of Qiskit. A basic Qiskit process consists of two stages: Build and Execute. Build allows you to create various quantum circuits that describe the problem at hand, and execution allows you to run them on various backends. Following the completion of the jobs, the data is gathered and post-processed based on the desired result. For example, the implementation of single-qubit gates in Qiskit is shown in Figure 12.2. The X gate rotates the state vector by π radians about the x-axis on the Bloch sphere, and it is represented by the matrix. An X gate has the syntax

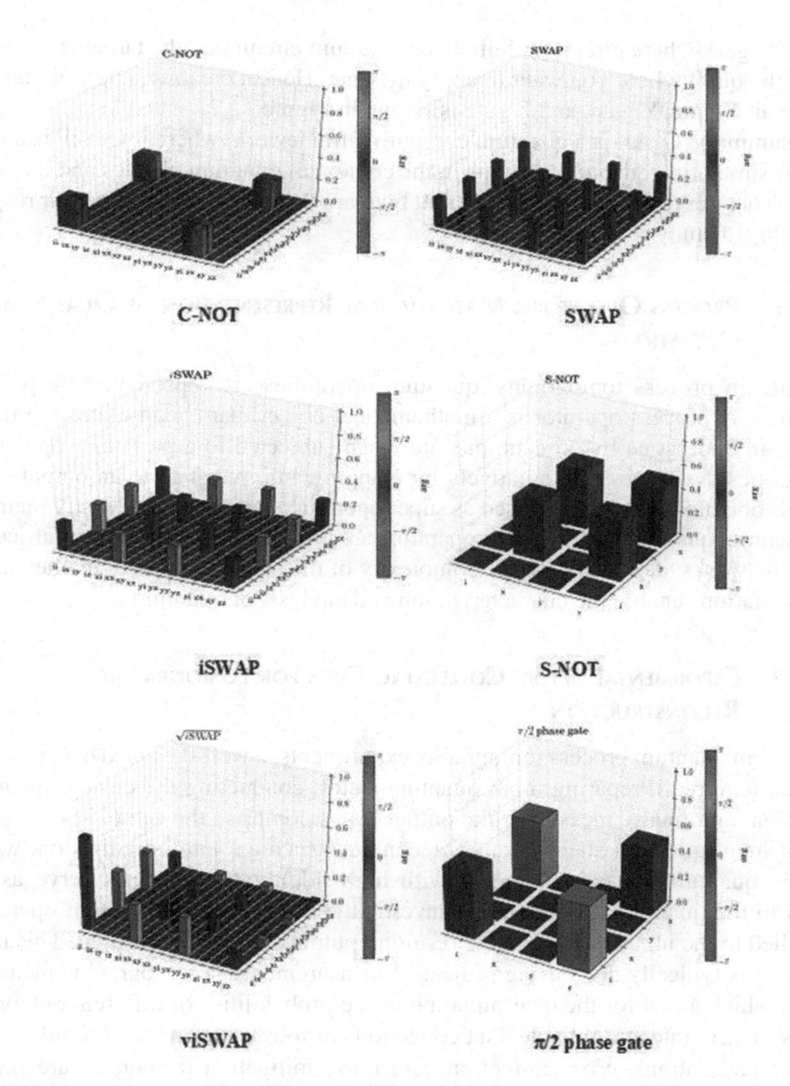

FIGURE 12.1 Quantum process tomography matrices for various qubit gates.

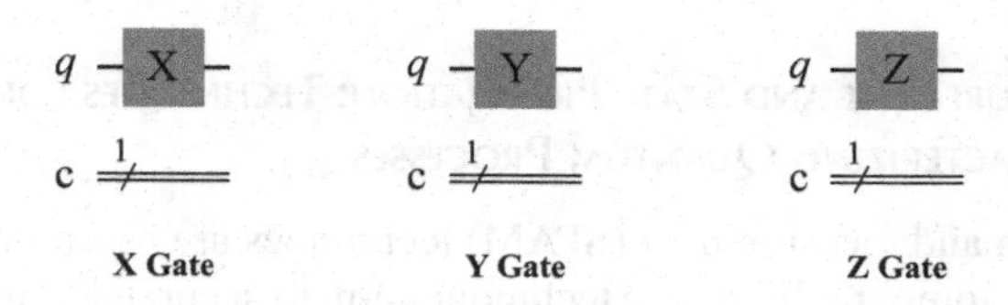

FIGURE 12.2 Single qubit implementation of X gate, Y gate, and Z gate.

circ. x(target), where circ. is an initialized quantum circuit and the target is the number of the qubit where you want to apply the gate. Hence, creating a new circuit and adding an X gate, Y gate, and Z gate is shown in Figure 12.2.

In summary, Qiskit is a quantum computing framework, whereas QuTiP is a toolbox for simulating quantum systems in the context of quantum physics and quantum optics. They serve different functions, but both are valuable resources in their respective fields of study.

12.2.1 Process Operators: Mathematical Representation of Quantum Operations

In quantum process tomography, quantum operations are represented by process operators. A process operator is a mathematical object that encapsulates the transformation undergone by a quantum state when subjected to a particular operation. It describes the process quantitatively by mapping an input state to an output state. Process operators can be expressed as super operators, which act on density matrices representing quantum states. Super operators are typically represented by matrices or tensor networks, depending on the complexity of the operation. These mathematical representations enable the characterization and analysis of quantum processes.

12.2.2 Experimental Setup: Collecting Data for Tomographic Reconstruction

To perform quantum process tomography experiments, a well-designed experimental setup is required. Preparing input quantum states, conducting the desired quantum operation, and finally measuring the output states comprise the setup. The preparation of input quantum states involves techniques such as state initialization, where specific quantum states are created with high fidelity. These states serve as the inputs to the quantum process under investigation. The desired quantum operation is applied to the input states, and the resulting output states are measured. This measurement is typically done using projective measurements or tomographic measurements, which allow for the determination of the probabilities of different outcomes. To ensure accurate and reliable data collection, various experimental considerations such as calibrations, error correction, and noise mitigation techniques are implemented. These measures help to minimize the impact of imperfections and errors in the experimental setup, ensuring the consistency of the obtained results.

12.2.3 Measurement and State Preparation: Techniques for Characterizing Quantum Processes

State preparation and measurement (SPAM) techniques are essential components of quantum process tomography. These techniques aim to accurately prepare the desired quantum states and perform precise measurements on the output states, enabling the characterization of quantum processes. State preparation techniques involve methods such as quantum state engineering, which allows for the controlled creation of specific quantum states. Techniques like quantum gates and pulse shaping enable the preparation of desired superposition states, entangled states, or other target states necessary for the tomographic reconstruction.

Measurement techniques play a crucial role in extracting information about the output states. Projective measurements, where the quantum system is projected onto a specific measurement basis, provide partial information about the state. Alternatively, tomographic measurements involve performing measurements in multiple bases to obtain a more complete description of the state. Both state preparation and measurement techniques should be carefully designed and calibrated to ensure accuracy and reliability. The success of quantum process tomography heavily relies on the ability to prepare desired input states and accurately measure the corresponding output states. Hence, quantum process tomography involves the mathematical representation of quantum operations, experimental setups for data collection, and the use of state preparation and measurement techniques. These methodologies aim to understand and characterize quantum processes, providing valuable insights into the behavior and quality of quantum operations.

12.3 GROUPS, COMPACT LIE GROUPS, AND THE UNITARY GROUP

Group theory is a mathematical framework that is essential for understanding and analyzing quantum physics symmetries. It provides sophisticated tools for analyzing quantum system features like as transformations, states, and observables. This section explores the fundamental concepts of group theory concerning quantum physics.

12.3.1 Group Theory: Fundamental Concepts

The group is a mathematical structure made up of a set of components and an operation that connects any two members in the set. In the context of quantum physics, groups capture the symmetries of the physical system. Symmetries are transformations that leave certain properties of the system unchanged, such as the form of the equations or the physical observables. Some key concepts in group theory include:

Group elements: The elements of a group are the individual entities within the set. In quantum physics, these elements may represent transformations, states, or other relevant entities.

Group operation: The group operation defines how the elements of a group are combined. It follows specific rules, such as closure (the result of the operation remains within the group), associativity (the order of operations does not matter), and the existence of an identity element (an element that, when combined with any other element, leaves it unchanged).

Group properties: Groups can have various properties, including commutativity (the order of operations does not affect the result), inverses (each element has a unique inverse that, when combined, gives the identity element), and the existence of a finite or infinite number of elements.

Subgroups: Subgroups are subsets of a group that possess the same group structure. They capture specific symmetries within a larger group.

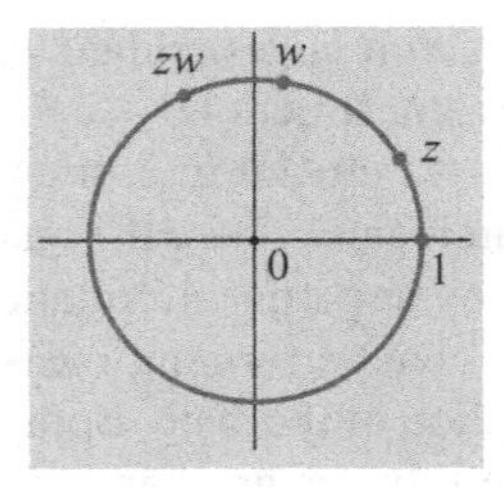

FIGURE 12.3 Compact Lie group.

12.3.2 COMPACT LIE GROUPS: STRUCTURE AND PROPERTIES

Compact Lie groups are a specific class of groups that have both compact and smooth properties. Compactness implies that the group is finite or bounded, while smoothness implies that the group has a well-defined differential structure. Compact Lie groups are important in quantum physics because of their rich mathematical features and tight relationship to symmetry. In quantum physics, symmetries are often represented by unitary transformations, and compact Lie groups provide a mathematical framework to describe these transformations. The unitary group (U(n)), rotation group (SO(3)), and the special unitary group (SU(n)) are all examples of compact Lie groups. In the complex plane, the circle with center 0 and radius 1 is a compact Lie group with complex multiplication as shown in Figure 12.3.

12.3.3 THE UNITARY GROUP: SIGNIFICANCE IN QUANTUM PROCESS TOMOGRAPHY

The set of unitary matrices of dimension $n \times n$ is represented by the unitary group, abbreviated as U(n). Unitary matrices are crucial in quantum process tomography because they preserve the norm and inner product of quantum states. This property ensures that probabilities are conserved during quantum operations, making unitary transformations fundamental in the processing of quantum information. In terms of quantum process tomography, the unitary group is significant because it provides a natural framework for describing and characterizing quantum operations. By representing quantum channels as unitary matrices, one can leverage the mathematical properties of the unitary group to study and analyze the behavior of quantum processes. Furthermore, the unitary group enables the application of quantum process tomography techniques, such as state preparation, unitary gates, and measurement schemes. These techniques, relying on the properties of unitary operations, allow for the accurate preparation of input states, the implementation of desired quantum transformations, and the measurement of output states. Hence, group theory provides a powerful mathematical framework for studying symmetries in quantum physics. Compact Lie groups, such as the unitary group, are particularly important due to their rich mathematical structure and connection to symmetries. Understanding the fundamental concepts of group theory and the significance of the unitary group is essential for analyzing quantum processes and performing quantum process tomography.

12.4 REPRESENTATION THEORY

Representation theory is the study of how abstract algebraic structures, such as graphs, can be represented. Groups, for example, are represented by linear transformations. In the setting of quantum physics, representation theory provides an effective framework for comprehending the relationship between group theory and quantum mechanics. This section introduces the basics and definitions of representation theory.

12.4.1 INTRODUCTION TO REPRESENTATION THEORY

Representation theory deals with the concept of representation, which is a way of associating each group element with a linear transformation of a vector space. The vector space on which the transformations act is known as the representation space. By studying representations, we can understand the group structure and its behavior. The study of how algebraic structures "act" on objects is known as representation theory. A basic example is how regular polygon symmetries, consisting of reflections and rotations, modify the polygon as shown in Figure 12.4.

Key definitions in representation theory include:

***Group representation*:** A group representation is a mapping from a group to a set of linear transformations of a vector space. It associates each group element with a specific linear transformation. Each representation of a group can be defined on a distinct vector space.

***Homomorphism*:** A homomorphism is a structure-preserving mapping between two mathematical structures. In representation theory, a group homomorphism maps group elements to linear transformations, ensuring that the group operation is preserved.

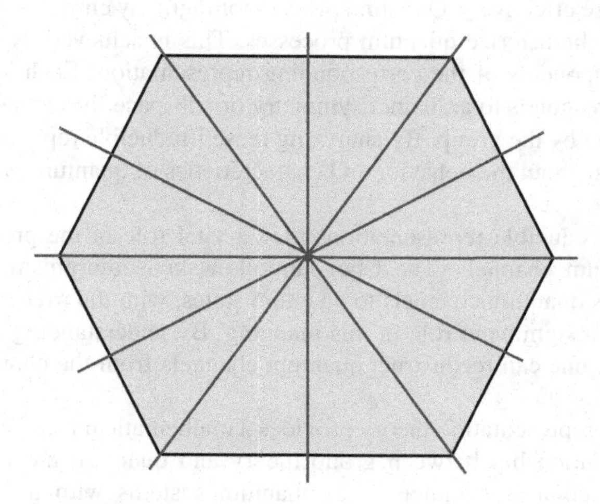

FIGURE 12.4 Regular polygon example for representation theory.

***Irreducible representation*:** This one cannot be split into smaller representations. It denotes the absence of non-trivial subspaces in the linked vector space that are invariant under the group's action.

***Reducible representation*:** A reducible representation can be decomposed into multiple irreducible representations. The associated vector space has non-trivial subspaces that are invariant under the group action.

12.4.2 Unitary Representations: Connecting Group Theory and Quantum Mechanics

Unitary representations are of particular importance in quantum mechanics because they preserve the inner product of vectors, ensuring the conservation of probabilities. A unitary representation is a representation in which the linear transformations are unitary operators. These unitary operators preserve the norm of vectors and conserve the inner product, reflecting the fundamental principles of quantum mechanics. Unitary representations connect the concepts of group theory and quantum mechanics by providing a mathematical framework for describing symmetries in quantum systems. Quantum states are commonly depicted as vectors within a Hilbert space, while the changes or operations performed in these states are symbolized by unitary operators. The unitary group, which consists of all unitary operators, plays a significant role in quantum process tomography and other quantum information processing tasks.

12.4.3 Irreducible Representations: Building Blocks for Quantum Process Tomography

Irreducible representations serve as building blocks in the study of quantum process tomography. By decomposing a representation into its irreducible components, one can understand the underlying structure of the quantum system and analyze its symmetries more effectively. Quantum process tomography employs irreducible representations to characterize quantum processes. This is achieved by examining the irreducible components of the corresponding representation. Each irreducible representation corresponds to a distinct symmetry or subspace that remains unchanged when acted upon by the group. By analyzing these irreducible representations, valuable information about the behavior and characteristics of quantum processes can be obtained.

Moreover, irreducible representations play a vital role in the process of reconstructing quantum channels. The Choi-Jamiolkowski isomorphism, mentioned in 12.1.3, connects quantum channels to quantum states, with the irreducible representations playing a significant role in this mapping. By understanding the irreducible representations, one can reconstruct quantum channels from the characterization of the corresponding states.

In summary, representation theory provides a mathematical framework for understanding the relationship between group theory and quantum mechanics. Unitary representations connect symmetries in quantum systems with unitary operators, reflecting the principles of quantum mechanics. Irreducible representations serve as fundamental components for analyzing quantum processes and reconstructing quantum channels in quantum process tomography.

12.5 STORAGE OF UNITARY AND PARALLEL APPLICATION

By utilizing the parallelism inherent in quantum systems, quantum computing has the potential for exponential speedup over classical computers. This section delves into the concept of parallelism in quantum computing, with a focus on quantum process tomography (QPT).

12.5.1 Parallel Application of Unitary Operators: Expediting QPT

In QPT, the characterization of quantum processes often involves the application of unitary operators on quantum states. Quantum computers can exploit the parallelism of qubits to accelerate the application of these unitary operators. In a classical computer, the application of a unitary operator on a state would involve a sequential execution of operations. However, in a quantum computer, multiple qubits can be in a superposition, allowing for the parallel application of unitary operators. This parallelism enables the execution of multiple operations simultaneously, potentially speeding up the process of QPT. By leveraging parallelism, quantum computers can process larger amounts of data and perform more complex calculations efficiently. This capability is particularly advantageous in QPT, where the reconstruction of quantum processes involves numerous unitary operations on quantum states.

12.5.2 Unitary Storage: Efficiently Encoding Unitary Matrices

Efficient encoding and storing of unitary matrices are crucial for implementing quantum process tomography. Unitary matrices represent quantum processes and their associated transformations accurately. Quantum computers can efficiently encode and manipulate unitary matrices due to their inherent parallelism. Representing and manipulating unitary matrices on a traditional computer can be computationally and memory-intensive. However, the parallelism of qubits in a quantum computer allows for more efficient storage and processing of unitary matrices. The intrinsic characteristics of quantum systems, such as entanglement and superposition, allow for more compact and concise representations of unitary matrices. Quantum computers utilize techniques such as quantum gates and quantum circuits to implement unitary transformations. These techniques leverage the parallelism of qubits to execute multiple unitary operations simultaneously, further enhancing the efficiency of quantum process tomography. Efficient unitary storage and manipulation are critical for practical quantum process tomography, as they directly impact the computational resources required and the accuracy of the reconstructed quantum processes.

In summary, parallelism in quantum computing provides significant advantages in quantum process tomography. By leveraging the parallel application of unitary operators and efficient encoding of unitary matrices, quantum computers can expedite the characterization of quantum processes. This parallelism allows for the efficient execution of operations, handling larger datasets, and performing complex calculations, ultimately pushing the boundaries of quantum process tomography.

12.6 OPTIMAL STATE FOR LEARNING

Within the realm of quantum machine learning, defining the optimal state is crucial for achieving effective learning and enhancing computational capabilities. This section explores the criteria used to determine the optimal state for quantum machine learning tasks.

12.6.1 Criteria for Effective Learning

Discriminative power: The optimal state should possess high discriminative power, meaning it can effectively distinguish between different classes or patterns. It should exhibit distinct features that enable accurate classification or regression tasks. Maximizing the discriminative power of the state enhances the learning accuracy and performance of quantum machine learning algorithms.

Entanglement: Entanglement, a phenomenon unique to quantum systems, is an essential criterion for effective learning. The optimal state should exhibit entanglement among its constituent qubits, as entanglement enables quantum systems to process and represent complex correlations between variables. Entangled states have the potential to provide exponential computational advantages over classical approaches.

Robustness to noise: Quantum systems are susceptible to noise and environmental disturbances, which can introduce errors in the learning process. The optimal state should be robust against noise, maintaining its discriminative power and information content even in the presence of perturbations. Ensuring the resilience of states against noise is imperative for practical implementations of quantum machine learning.

12.6.2 Quantum Machine Learning (QML) Algorithms: Leveraging Quantum States

Quantum machine learning (QML) algorithms leverage the unique properties of quantum states to enhance learning capabilities. These algorithms exploit the parallelism, superposition, and entanglement of quantum systems to perform tasks such as classification, clustering, and regression more efficiently compared to classical approaches. By utilizing quantum states as computational resources, QML algorithms are capable of processing and manipulating massive volumes of data in parallel. The property of superposition exhibited by quantum states enables the simultaneous exploration of multiple hypotheses or solutions, enhancing computational efficiency. Quantum support vector machines, quantum variational algorithms, and quantum neural networks are examples of QML algorithms that use quantum states to optimize learning tasks.

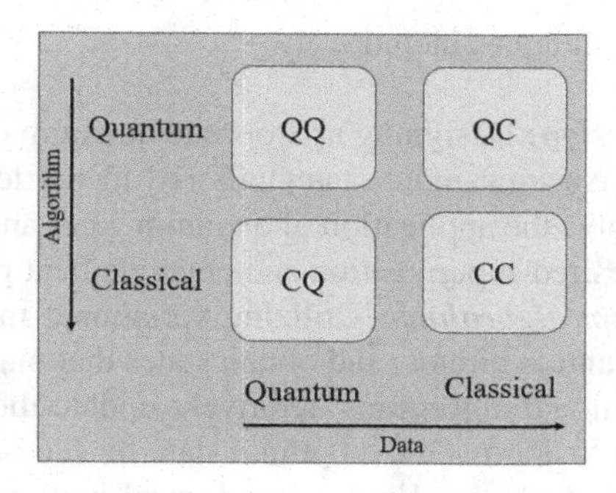

FIGURE 12.5 Quantum machine learning categories.

These algorithms utilize quantum gates and quantum circuits to manipulate quantum states, extract features, and optimize parameters for effective learning. For example, because of its remarkable superposition and entanglement capabilities, quantum machine learning (QML) has attracted substantial attention in disciplines such as medical image analysis, password cracking, and pattern recognition. Despite its potential, traditional machine learning confronts limitations such as a lack of labeled data and limited processing efficiency. The combination of quantum computing and machine learning has increased parameter optimization, execution efficiency, and error rates. In terms of speed and performance, QML algorithms beat standard machine learning methods. Furthermore, QML has tremendous computing capabilities and is currently being developed by researchers all around the world. The current focus of research is on speeding up classical machine learning algorithms and implementing quantum algorithms in classical computers. However, because of an absence of class I algorithms and a scarcity of hardware, QML is difficult to create and implement. Future research should concentrate on developing more efficient encoding methods and investigating QML algorithms in real-world applications such as natural language processing, speech recognition, and recommendation systems. QML takes advantage of quantum computing's engaged parallelism to improve classical machine learning. As shown in Figure 12.5, machine learning can be classified based on the algorithm's form (classical or quantum) and the pattern (classical or quantum) of the processed data.

It is classified into four categories, three of which (QQ, QC, and CQ) are termed QML. The first category of QML is the use of quantum algorithms on quantum computers (QQ), the second is the use of quantum computers to accelerate classical machine learning algorithms (quantum/classical hybrids, QC), and the third is the use of quantum-inspired algorithms on classical computers (CQ) (Wei et al. 2023).

12.6.3 State Engineering Techniques: Preparing Optimal Input States

State engineering techniques play a vital role in quantum machine learning by preparing optimal input states that enhance learning performance. These techniques involve the manipulation of quantum states to enhance discriminative power, promote entanglement, and optimize the quantum machine learning algorithm's efficiency.

State engineering techniques include:

***Quantum circuit design*:** Designing appropriate quantum circuits or networks that can prepare specific input states tailored to the learning task. These circuits may involve the application of quantum gates and entangling operations to create desired superposition and entanglement patterns.

***Variational quantum algorithms*:** Utilizing variational methods to optimize parameters in quantum circuits and obtain states that maximize the learning objectives. Variational algorithms iteratively update the parameters of the quantum circuits to approach the optimal state for the given learning task.

***Quantum control techniques*:** Employing control methods, such as optimal control theory or machine learning-inspired control strategies, to manipulate quantum systems and engineer desired input states. These techniques optimize the control parameters to shape the quantum states' properties according to the learning requirements.

By employing state engineering techniques, quantum machine learning practitioners can design and prepare input states that meet the criteria for effective learning. These techniques allow for the customization and optimization of quantum states to enhance the performance and efficiency of QML algorithms. Hence, defining the optimal state in QML involves considering criteria such as discriminative power, entanglement, and robustness to noise. Quantum machine learning algorithms leverage quantum states and their unique properties to enhance learning capabilities. State engineering techniques play a crucial role in preparing optimal input states that maximize learning performance, enabling quantum systems to tackle complex learning tasks more effectively.

12.7 FINDING THE PARAMETER FOR THE INPUT STATE AND APPLYING THE UNITARY OPERATORS

In quantum information processing and quantum process tomography, the use of unitary operators is crucial in the transformation of quantum states. This section explores how unitary operators apply to quantum states and their significance in various applications. Some applications of unitary operators are discussed. Unitary operators represent reversible quantum transformations that preserve the norm and inner product of quantum states. They are fundamental in quantum process tomography as they capture the behavior of quantum channels or processes. The application of a unitary operator on a quantum state involves multiplying the state vector by the corresponding unitary matrix. This transformation changes the state's representation in the underlying Hilbert space, encoding the desired quantum operation. The resulting state represents the output after the application of the unitary operator. Unitary operators are used in a variety of quantum information processing activities, including quantum gates for quantum computing, quantum algorithms, and quantum error correction. By carefully designing and implementing unitary operators, quantum states can be manipulated to perform specific computing or communication tasks.

12.7.1 Parameter Estimation: Techniques for Determining Input State Parameters

In quantum process tomography, accurately characterizing quantum processes requires determining the parameters of the input quantum states. Parameter estimation techniques play a crucial role in finding the optimal values of these parameters to obtain desired output states. Parameter estimation involves extracting information about the input state by performing measurements on the output state after applying the unitary operator. By comparing the measured data with the expected outcomes, estimation algorithms can infer the values of the input state parameters. Various techniques are employed for parameter estimation, including maximum likelihood estimation, Bayesian estimation, and quantum state tomography. These techniques leverage statistical analysis and optimization algorithms to determine the parameters that best match the observed data.

12.7.2 Optimization Algorithms: Finding the Best Input State

Optimization algorithms are utilized to find the best input state for a given quantum process tomography task. These algorithms search for the optimal values of input state parameters that maximize the desired objective, such as fidelity, accuracy, or information gain. Different optimization algorithms can be employed, including genetic algorithms, gradient-based methods (Ahmed et al. 2023), simulated annealing, or variational methods. These algorithms iteratively update the input state parameters based on the feedback obtained from the measurement outcomes and the optimization objective. The choice of optimization algorithm depends on the specific problem and its associated constraints. Factors such as computational complexity, convergence properties, and robustness to noise influence the selection of the appropriate algorithm. Optimization algorithms are crucial for obtaining high-quality reconstructions of quantum processes. They help identify the input state parameters that lead to optimal learning outcomes, enabling accurate characterization and understanding of the underlying quantum processes. Hence, the application of unitary operators transforms quantum states in quantum process tomography and other quantum information processing tasks. Parameter estimation techniques play a vital role in determining the input state parameters based on measured data. Optimization algorithms aid in finding the best input state by optimizing objective functions. Together, these components contribute to accurate characterization and efficient utilization of quantum processes.

12.8 SUMMARY

12.8.1 Recapitulating Quantum Process Tomography and Regression

Quantum process tomography and regression are powerful techniques that enable the characterization and understanding of quantum processes. This chapter has covered a variety of key concepts and strategies connected to QPT and regression. To summarize the important points discussed: Quantum process tomography focuses on reconstructing and characterizing quantum channels or processes. It involves the mathematical representation of quantum operations, experimental setups for data collection, state preparation and measurement techniques, and the use of representation theory and group theory to analyze and understand quantum processes. The chapter delved into the concepts of channel-state duality, highlighting the deep relationship between quantum channels and states. The duality reveals that the transformation of the operation of a channel in a quantum state can be characterized in the same way. This connection forms the basis for quantum process tomography and allows for the reconstruction of quantum channels by characterizing the associated states. Representation theory and group theory play fundamental roles in quantum process tomography. Compact Lie groups, such as the unitary group, provide a mathematical framework for describing symmetries and transformations in quantum systems. Irreducible representations serve as building blocks for analyzing quantum processes and reconstructing quantum channels. Parallelism in quantum computing was explored as a means of leveraging multiple qubits to expedite quantum process tomography. The parallel application of unitary operators and efficient encoding of unitary matrices enable more efficient computations and data processing in quantum machine learning. State engineering techniques were discussed for preparing optimal input states in quantum process tomography. By manipulating quantum states, one can enhance discriminative power, promote entanglement, and optimize the efficiency of quantum machine learning algorithms. Parameter estimation techniques and optimization algorithms were explored as essential tools for determining input state parameters and finding the best input states. These techniques enable the accurate characterization of quantum processes and facilitate effective learning.

12.8.2 Implications and Future Directions in Quantum Computing and Quantum Information Processing

The advancements in quantum process tomography and regression have profound implications for quantum computing and quantum information processing. They pave the way for better understanding and utilization of quantum systems. Some implications and future directions include:

***Quantum algorithm design*:** Quantum process tomography helps in designing efficient quantum algorithms by characterizing the behavior of quantum processes. This understanding can be used to increase the efficiency and precision of quantum algorithms in a range of applications.

***Quantum error correction*:** Quantum process tomography assists in characterizing and mitigating errors in quantum systems. It provides insights into the noise sources, allows for error correction schemes, and aids in the development of fault-tolerant quantum computing architectures.

***Quantum machine learning*:** The combination of regression and quantum process tomography with machine learning techniques holds great potential for advancing quantum machine learning algorithms. By characterizing quantum processes accurately, one can optimize learning performance and leverage quantum advantages for pattern recognition and data analysis tasks.

Quantum communication and cryptography: Understanding quantum processes is crucial for secure communication and cryptography protocols. Quantum process tomography helps in characterizing the behavior of quantum channels, ensuring reliable and secure transmission of quantum information.

Quantum control: The knowledge gained from quantum process tomography can be applied to quantum control techniques, enabling precise manipulation and engineering of quantum systems. This control can lead to improvements in quantum information processing tasks and the development of advanced quantum technologies.

Quantum simulators offer significant advantages over classical simulations in specific computational tasks due to their ability to simulate quantum systems more efficiently as shown in Figure 12.6. These advantages include exponential speedup, quantum parallelism, and the ability to model complex quantum systems. Quantum simulators can simulate large quantum systems in a time complexity that scales exponentially with system size, allowing for more accurate results. They excel in modeling and understanding quantum chemistry, condensed matter physics, and high-energy physics, providing more precise insights into quantum phenomena. Quantum simulators also have a quantum advantage in specific applications, such as drug discovery and materials science. However, they face challenges like quantum noise and errors, which are being addressed through quantum error correction techniques. As quantum hardware advances, the quantum advantage of quantum simulators is expected to grow, making larger and more complex quantum systems more feasible to simulate.

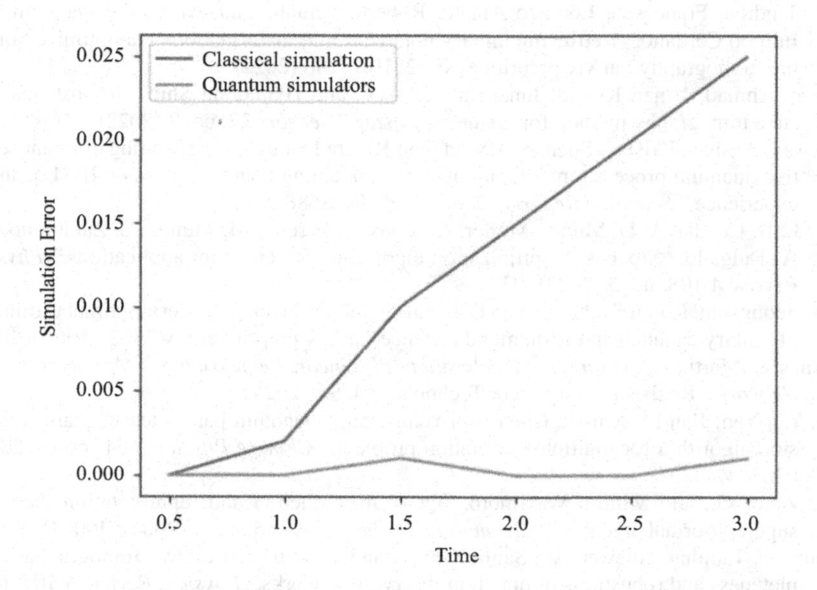

FIGURE 12.6 Quantum simulation over classical simulation.

In summary, quantum process tomography and regression are powerful tools for characterizing, understanding, and utilizing quantum processes. They provide insights into the behavior of quantum channels, enable efficient quantum information processing, and open doors for advancements in quantum computing, quantum machine learning, quantum communication, and quantum control

REFERENCES

Ahmed, Shahnawaz, Fernando Quijandría, and Anton Frisk Kockum. "Gradient-descent quantum process tomography by learning Kraus operators." *Physical Review Letters* 130, no. 15 (2023): 150402.

Wei, Lin, Haowen Liu, Jing Xu, Lei Shi, Zheng Shan, Bo Zhao, and Yufei Gao. "Quantum machine learning in medical image analysis: A survey." *Neurocomputing* 525 (2023): 42–53.

FURTHER READING

Arunachalam, Srinivasan, Sergey Bravyi, Arkopal Dutt, and Theodore J. Yoder. "Optimal algorithms for learning quantum phase states." arXiv preprint arXiv:2208.07851 (2022).

Bisio, Alessandro, Giulio Chiribella, Giacomo Mauro D'Ariano, Stefano Facchini, and Paolo Perinotti. "Optimal quantum learning of a unitary transformation." *Physical Review A* 81, no. 3 (2010): 032324.

Cheng, Yanran, and Zhihui Lou. "A brief review of linear regression estimation in quantum tomography." In *2020 39th Chinese Control Conference (CCC)*, pp. 5813–5817. IEEE, 2020.

Daley, Andrew J., Immanuel Bloch, Christian Kokail, Stuart Flannigan, Natalie Pearson, Matthias Troyer, and Peter Zoller. "Practical quantum advantage in quantum simulation." *Nature* 607, no. 7920 (2022): 667–676.

Di Colandrea, Francesco, Lorenzo Amato, Roberto Schiattarella, Alexandre Dauphin, and Filippo Cardano. "Retrieving unitary polarization transformations via optimized quantum tomography." arXiv preprint arXiv:2210.17288 (2022).

Farooq, Ahmad, Uman Khalid, Junaid ur Rehman, and Hyundong Shin. "Robust quantum state tomography method for quantum sensing." *Sensors* 22, no. 7 (2022): 2669.

Gaikwad, Akshay, Krishna Shende, Arvind, and Kavita Dorai. "Implementing efficient selective quantum process tomography of superconducting quantum gates on IBM quantum experience." *Scientific Reports* 12, no. 1 (2022): 3688.

Gidi, J., B. Candia, A. D. Muñoz-Moller, A. Rojas, L. Pereira, M. Muñoz, L. Zambrano, and A. Delgado. "Stochastic optimization algorithms for quantum applications." *Physical Review A* 108, no. 3 (2023): 032409.

Haah, Jeongwan, Robin Kothari, Ryan O'Donnell, and Ewin Tang. "Query-optimal estimation of unitary channels in the diamond distance." arXiv preprint arXiv:2302.14066 (2023).

Hoffnagle, Martin A. *Quantum Acceleration of Linear Regression for Artificial Neural Networks*. Rochester Institute of Technology, USA, 2023.

Hou, Yan-Yan, Jian Li, Xiu-Bo Chen, and Yuan Tian. "Quantum partial least squares regression algorithm for multiple correlation problem." *Chinese Physics B* 31, no. 3 (2022): 030304.

Kac, Victor G., and Minoru Wakimoto. "Quantum reduction and representation theory of superconformal algebras." *Advances in Mathematics* 185, no. 2 (2004): 400–458.

Krisnanda, Tanjung, Huawen Xu, Sanjib Ghosh, and Timothy CH Liew. "Tomographic completeness and robustness of quantum reservoir networks." *Physical Review A* 107, no. 4 (2023): 042402.

Laredo, Valerio Toledano, and Xiaomeng Xu. "Stokes phenomena, Poisson–Lie groups and quantum groups." *Advances in Mathematics* 429 (2023): 109189.

Mu, Biqiang, Hongsheng Qi, Ian R. Petersen, and Guodong Shi. "Quantum tomography by regularized linear regressions." *Automatica* 114 (2020): 108837.

Ning, Tong, Youlong Yang, and Zhenye Du. "Quantum kernel logistic regression based Newton method." *Physica A: Statistical Mechanics and its Applications* 611 (2023): 128454.

Ragone, Michael, Paolo Braccia, Quynh T. Nguyen, Louis Schatzki, Patrick J. Coles, Frederic Sauvage, Martin Larocca, and M. Cerezo. "Representation theory for geometric quantum machine learning." arXiv preprint arXiv:2210.07980 (2022).

Saki, Abdullah Ash, Mahabubul Alam, and Swaroop Ghosh. "Quantum true random number generator." In *Design Automation of Quantum Computers*, pp. 69–86. Cham: Springer International Publishing, 2022.

Surawy-Stepney, Trystan, Jonas Kahn, Richard Kueng, and Madalin Guta. "Projected least-squares quantum process tomography." *Quantum* 6 (2022): 844.

Verdeil, François, and Yannick Deville. "Two unitary quantum process tomography algorithms robust to systematic errors." *Physical Sciences Forum* 5, no. 1 (2022): 29. MDPI.

Xiao, Shuixin, Yuanlong Wang, Daoyi Dong, and Jun Zhang. "Optimal and two-step adaptive quantum detector tomography." *Automatica* 141 (2022): 110296.

Yan, Bin, and Nikolai A. Sinitsyn. "Randomized channel-state duality." arXiv preprint arXiv: 2210.03723 (2022).

Yu, Qi, Daoyi Dong, Yuanlong Wang, and Ian R. Petersen. "Adaptive quantum process tomography via linear regression estimation." In *2020 IEEE International Conference on Systems, Man, and Cybernetics (SMC)*, pp. 4173–4178. IEEE, 2020.

Zhang, Mao, Huai-Ming Yu, Haidong Yuan, Xiaoguang Wang, Rafał Demkowicz-Dobrzański, and Jing Liu. "QuanEstimation: An open-source toolkit for quantum parameter estimation." *Physical Review Research* 4, no. 4 (2022): 043057.

Zia, Danilo, Riccardo Checchinato, Alessia Suprano, Taira Giordani, Emanuele Polino, Luca Innocenti, Alessandro Ferraro, Mauro Paternostro, Nicolò Spagnolo, and Fabio Sciarrino. "Regression of high-dimensional angular momentum states of light." *Physical Review Research* 5, no. 1 (2023): 013142.

Index

Pages in *italics* refer to figures

D

E

J

K

L

M

N

O

P

Q

R

S

T

U

V

W

X

Y

Z

Printed in the United States
by Baker & Taylor Publisher Services

Printed in the United States
by Baker & Taylor Publisher Services